3

Wheeler & Wilson
'A Stitch in Time'

By
Alex Askaroff

Sewing Machine
Pioneer Series

Wheeler & Wilson
'A Stitch in Time'

By
Alex Askaroff

Sewing Machine
Pioneer Series

To see Alex Askaroff's work
Visit Amazon
&
www.sewalot.com

The rights of Alex Askaroff as author of this work have been asserted by him in accordance with the Copyright, Designs and Patents Act 1993.
©

ISBN: 9781709364204

This is no masterpiece. It's more a self-published labour of love from someone who has spent a lifetime in the sewing trade and a million hours gathering facts for you. From my vast profits of around sixpence a book I'll buy a new tyre for my bicycle and ride to my allotment to tend my cabbages. Why write it? Well no one else bothers! Please forgive my spelling, United Kingdom English, and enjoy it in the spirit it was written.

Wheeler & Wilson
A Stitch in Time

The brass badge with the Wheeler & Wilson shield was inlaid into many of their machines.

The Wheeler & Wilson Manufacturing Company Bridgeport, Connecticut,

Makers of the finest sewing machines in the world.

A rare cover showing the famous Wheeler & Wilson emblem and Trademark. This cover is now in my Sewalot Collection

Index

Introduction..8

Chapter One..10

Chapter Two..14

Chapter Three...24

Chapter Four...33

Chapter Five..44

Chapter Six...50

Chapter Seven..51

Chapter Eight..54

Chapter Nine...61

Chapter Ten..64

Chapter Eleven.......................................73

Chapter Twelve......................................76

Chapter Thirteen....................................81

Chapter Fourteen...................................90

Chapter Fifteen.....................................108

Chapter Sixteen....................................115

On The Road Series

There are seven books in Alex Askaroff's **On The Road Series**. They cover his working life encompassing a world of sewing stories from the ages.

Book One: Patches of Heaven

Book Two: Skylark Country

Book Three: High Streets & Hedgerows

Book Four: Tales From The Coast

Book Five: Have I Got A Story For You

Book Six: Glory Days

Book Seven: Off The Beaten Track

"If you read any of Alex's 'On The Road Series' you will read them all. They are totally addictive, beautifully crafted and wonderfully inspiring."
Eliza Cooper

Introduction

Wheeler & Wilson are names we hardly hear today, but once they were the largest sewing machine manufacturers in the world. The massive factories in Bridgeport, Connecticut, covered acres of ground. The chimneys blackened the air with smoke pouring from the forges that were working hot metal for the new era of invention. Nathaniel Wheeler and Allen Benjamin Wilson were vastly different characters. One was strong, forceful and full of energy. The other weak but with a brilliant, inventive mind.

At its height Wheeler & Wilson had huge factories and employed thousands of workers. This is their plant, which actually was not on Broadway as the building seems to imply. To my knowledge there were no Wheeler &Wilson factories on Broadway. Their main factory was initially in Watertown, then Bridgeport, Connecticut. However, their sales and storage depots were mainly on Broadway & Union Square, New York.

This research is the accumulation of a lifetime spent in the sewing trade. It is the most comprehensive archive yet published on Wheeler & Wilson.

How did it all happen? Our tale starts in Pittsfield, Massachusetts way back in 1841 with Allen Benjamin Wilson, an inventive 18 year old who tinkered with the idea of making his own sewing machine. An apprentice cabinet maker and journeyman (an apprentice that has completed his term as apprentice and paid a daily wage), he was a young man that yearned for something bigger and better out of life.

Plagued with poor health and a nervous disposition he would often have to stay at home. This gave him the time (during his illnesses) to build working models of his amazing ideas. Although Allen B Wilson had pretty much finished with the sewing business by the age of 30, the 12 years he spent in them changed our world.

This book will fall into mostly two sections: the first part will be on our main two characters, Allen Benjamin Wilson and Nathaniel Wheeler, and the second section will be more about the production of sewing machines, serial numbers and models. Hold onto your hats, here we go.

Chapter One

The Early life of A B. Wilson
18 Oct 1823 – 29 Apr 1888

The village of Willete (note the old spelling) was founded by six main men around 1806-7, one of them being Benjamin Willson Senior, Allen Benjamin Willson's grandfather, (note the two ll's in Willson at this time).

Benjamin Willson Sr emigrated from England. At the time of the War of Independence (1775-1783) he was a young Tory and sided with the British. He married Phebe and had 14 children. He went on to become one of the successful Founding Fathers of Willet.

In Willet, Benjamin Willson Sr set up several enterprises, which included a distillery, a gristmill, sawmill, ashery, Inn, store and blacksmith shop. Though siding with the British when hostilities broke out, through strength of character, loyalty to his friends and determination, he became one of the most important men of the area.

His son, Benjamin Willson Junior, married Francis (Fanny) on 17 October 1817. Ben's wife went on to have several children, one of whom was Allen B. Willson, inventor and sewing machine pioneer. There was also Phebe (named after his mother) and Mary.

Allen B. Willson was born in the village of Willet (Willette), NY on 18 October 1823.

However, on 18 January 1826, when Allen was just a child, his father was killed in a tragic accident at his grandfather's mill. Allen's grandfather stepped in to help but, when Allen was just 13 his grandfather also died.

Benjamin Willson Junior, his wife, Francis and daughter Phebe are all buried in Willet.

Now let us go back to those early years in Willet NY and travel with Allen B. Wilson as he grows to become one of the great sewing machine pioneers of the 19th Century.

Allen Benjamin Willson becomes Allen B Wilson

There is little doubt that Allen Benjamin Willson would have grown up working around the various family businesses along with his mother and other siblings. This would have been the perfect grounding for an inventive mind, being surrounded by machinery of all shapes and sizes. The noise and movement of the smiths and mills were to shape Allen's life and help guide his inventive mind.

Allen moved to Adrian, Michigan and at the age of 27 he married Harriet Emeline Brooks (2 May 1829-14 Sept 1895) from Massachusetts. For some reason Allen B Willson then dropped the extra 'l' in his name and became Allen Benjamin Wilson with one l.

Allen and Harriet had two daughters, Annah Bennette Wilson and Harriet Ethel Wilson. Harriet's tale is a tragic one and she ended up in a poorhouse. The reasons and details have been lost in the mists of time and even in death the family were never reunited. Although a space was provided for her at the family plot, it was never used. How this came about is still a mystery.

Allen B. Wilson died on 29 April 1888 in Waterbury. He was buried with his wife

Harriet and daughter Annah in Waterbury CT.

This is a super rare image of Allen B Wilson and his wife Harriet. To take the image they had to sit perfectly still for ages. They were told not to smile as when the smile dropped it gave a blurred effect to the photo.

Chapter Two

Now, on a happier note, back to the story of Allen B Wilson and his part in the history of the sewing machine.

The 'Seaming Lathe' or A B. Wilson sewing machine

Allen B Wilson's first sewing machine ideas were nothing like the machines we see today. His feed mechanism was a basic bar that gripped the cloth and pulled it along.

It was his later improvements that were spectacular. So spectacular that many are still in use today.

In 1848, while Allen was working as a cabinetmaker in Pittsfield, Massachusetts the business suddenly closed. Out of work, Allen did a deal with its owner, Amos Barnes, which allowed him to use his workshop. In the workshop he could carry on with the idea that he had to build a machine he called his 'seaming or sewing lathe'.

By the spring of 1849 Allen had perfected enough of the movement to secure an investor, one Joseph Chapin. Joseph invested $211 in Allen's invention. This allowed him enough money to secure his first patents. By late 1849 (with his patent secured in England, Patent 12752) he applied for it in America. Was this Allen's way of checking that his ideas were original or was he planning ahead?

November 12, 1850 US Patent 7776

On 12 November 1850 (the year before Isaac Singer patented his machine) Allen B Wilson had begun his long journey in the sewing machine field with patent 7776 for sewing machine shuttle improvements. It was still only the 15th American patent concerning sewing machines!

Allen farmed out some of the early precision work on his first sewing machines to Deall & Sons. It was cheaper to get other engineering firms to make certain parts, rather than make the huge investment in the machinery needed to make them himself.

The Wilson patent model of 1850. It was soon replaced by the 'improved patent model'. It was one of the first proper sewing machines. The machine that went into mass production did not look like this beauty.

Patent No 8296
12th August 1851
Rotary Hook sewing machine with reciprocating bobbin
Patent 9041
15th June 1852

Rotary hook sewing machine with stationary bobbin
Granted to
A. B. Wilson

To avoid litigation, which was rife, Wilson was secretly designing a completely new form of sewing machine. Eventually it would contain three superb and unique ideas: the four-motion-feed, the rotary hook, and the stationary bobbin (inside the rotating hook).

Each time Wilson perfected another piece he patented it. This practice of patenting as many ideas as possible was because of the frenzy of court cases taking place in the 1850's. Remember, we are talking about the birth of the first sewing machine industry in the world.

Shortly after this the patents were applied for all over Europe.

His first shuttle was a weird double-pointed affair that produced two stitches with every back-and-forward movement. However from these early ideas he then produced several startling and innovative pieces of engineering. Within 24 months he had produced a world-beating machine.

The Improved Sewing Lathe 1852.
Casting this stunning model out of iron must have been a nightmare and it did not last long. It incorporated the eye-pointed needle patented by Elias Howe (although Wilson had to curve his needle to work). Royalties had to be paid to Elias Howe for any machine using his patented needle design.

The real deal. This stunning early 1850's Wheeler & Wilson came up for auction in Boston a few years back. Mike Anderson from Wolfegang's Collectibles was outbid and it went for $22,000 to one of the premier American collectors.

With Wilson's Patent 9041 of 1852, for the stationary bobbin secure, he officially christened this machine his Improved Seaming Lathe! Within a few years his machine had gone from a very flamboyant looking patent model to a more obvious and practical sewing machine. Amazingly the model below ran from the 1850's right up until the takeover of the Wheeler & Wilson factory in 1905. Besides the superb Wilcox & Gibbs chainstitch it had the longest production run of any 19th Century sewing machine.

Allen Benjamin Wilson
Oct 18, 1823 - April 29, 1888

Allen Benjamin Wilson in his prime. Kindly supplied by Helen DeFoe the distant Great Niece of A B. Wilson. Helen also provided some of the family history and the hotel picture. Here you can easily see the frail countenance of Allen who suffered terribly from his court battles.

This is the side view of the June 1852 patent for the Sewing Lathe. Wilson's machine contained all the necessary mechanisms to produce a fine lockstitch with two threads. Notice the large cylindrical barrel in the centre. That is the wooden pulley that the flat treadle belt would roll around to turn the sewing lathe. With this model Wilson had now entered the playing field for the biggest market on earth, that of the sewing machine.

Now you can start to see the machine that we are familiar with. This model was so popular it ran right up to the end of the 19th Century.

Unfortunately owners of the John Bradshaw shuttle patent quickly pounced on Wilson, halting his sewing machine manufacture. Kline & Lee pulled a cracker of a bluff on him. They told Allen that his patent would infringe on theirs, leading to him being prosecuted! Wilson was in no financial condition to take these patent holders to court, even if their patent was not that similar to his. By this time he had spent almost every penny of his $211 investment from Joseph Chapin on securing his patent.

He had little choice but to sign over half his patent to A. P. Kline & E. E. Lee. Interestingly he also went into business with them but it was an acrimonious arrangement built on distrust (and their claims over his patent).

Wilson struggled on with his machines, however a fortunate meeting back in the year 1850 was about to bear fruit. He had the great luck to bump into Nathaniel Wheeler, partner in the Warren, Wheeler & Woodruff Manufacturing Company of Watertown, Connecticut. This meeting would eventually lead to the Wheeler & Wilson Manufacturing Company being born.

The Wheeler & Wilson 'New' No1 or Y-Arm was a bestseller for the company. The model was outdated and out-performed by almost every other maker of the period but W&W had such a loyal following and undisputed quality that the demand was still there. At $50 it was a great looker as well. Interestingly, the operator sat to the side of the machine feeding the work left-to-right.

Here is a real raised bed Y-Arm sold as a light industrial by Wheeler & Wilson. Picture Mike Anderson.

1858

Allen Benjamin Wilson at the height of his fame in the sewing machine business. Kindly supplied by Helen DeFoe, the Great Niece of A B Wilson. Allen was 35 when this daguerreotype image was taken by the famous New York photographer, J Gurney. His studio was at 349 Broadway, New York. He had a select list of clients, mainly the rich and famous.

Chapter Three

There will be more of Nathaniel Wheeler in a moment, but first let us look at some of Wilson's amazing ideas during his brief period in the sewing machine business.

June 16, 1852
To all whom it may concern.
I, Allen B. Wilson of Watertown, in the county of Litchfield and the state of Connecticut, have invented certain new and useful improvements in the machinery for sewing.

Allen B Wilson's best ideas by far were the rotary hook mechanism and his brilliant 'four motion feed'. The rotary hook simply went round and round, in smooth endless circles. On its travels a point picked up the top thread from the needle, twisted it with a thread from the bobbin and let it go. So simple and smooth that it would last a lifetime. In fact it ended up lasting forever because many machines today still use his simple mechanism.

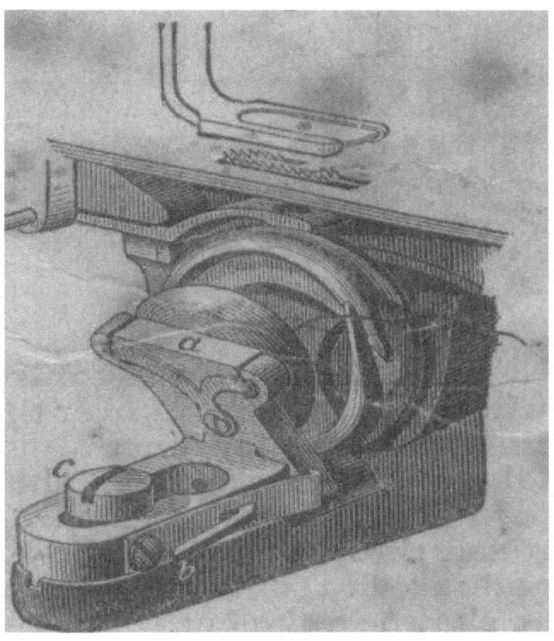

The rotary hook of Allen B. Wilson, note the brilliant idea of a glass see-through foot. It was only with the invention of clear plastic sewing feet a century later that see-through sewing feet returned. Notice how the bobbin case retainer hinges open. The first models did not have this so they gave you a screwdriver to undo bolt C and slide the retainer back. The machine was constantly under improvement.

Wilson's US Patent for the stationary bobbin Patent 9041 June 15, 1852

No sewing machine had used this type of idea before and seeing as how humans had been trying for centuries to figure out a perfect stitch, this really was genius. If you turn a Wheeler & Wilson sewing machine it is so light, smooth and quiet you would be forgiven for thinking it was made recently, not back at the dawn of the great sewing machine age.

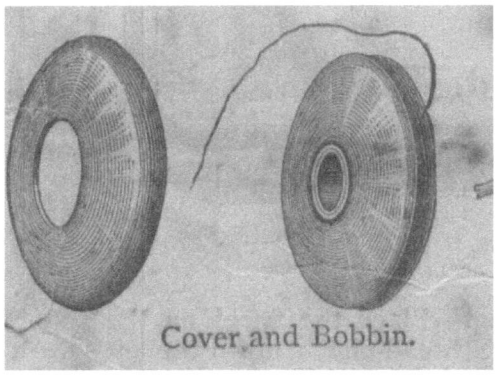
Cover and Bobbin.

The bobbin and bobbin cover sat under the machine and were unique to Wheeler & Wilson machines. It was an incredibly narrow bobbin. In Britain it was called the 'penny' bobbin as it was almost the same size.

His second stroke of genius is more in dispute. He needed a better method of moving the cloth through the machine than his early bar method. At the time when people were trying rows of pins, spiked wheels and other silly ideas, Allen came up with a set of forward leaning teeth. They appeared, as if by magic, from under the work, moved the work forward and then disappeared again.

People looked on in amazement at this 'black art'. It turned out to be the 'FOUR MOTION FEED', simple engineering brilliance, well ahead of its time. Or was it? Read on my friends, for there is a snag ahead.

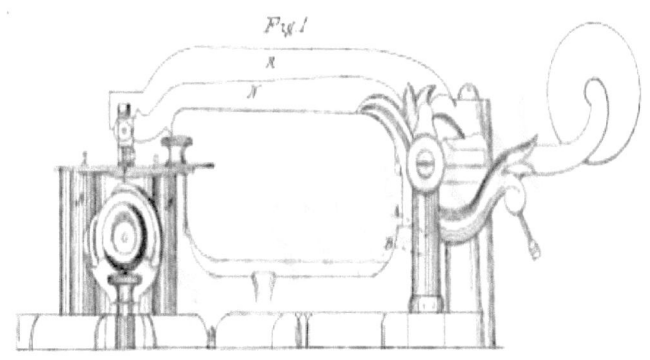

The A B Wilson patent of Dec 1854 No 12116
Allen B Wilson's 1854 patent improved machine was an elegant machine compared to Singer models of the same period. Plus it incorporated the 'four motion feed'.

Coincidentally the Grover & Baker sewing machine used a very similar method of moving the work forward. So who was first? Allen B. Wilson had already been cheated out of one of his earlier patents when Kline & Lee conned him into believing that his double pointed shuttle was really a copy of their 1848 Bradshaw Patent.

When Grover & Baker told him to cease and desist or be prosecuted, he was furious. This time he was ready for a fight.

In later life Allen B Wilson let his beard run wild. Hairy Mary! I wonder if he lost stuff in there! Maybe breakfast…

Earlier, Wilson had given up without much of a fight, allowing Kline & Lee to take half his patent rights for the shuttle, and eventually he relinquished the full rights to it. Good riddance to bad rubbish he must have thought.

Mind you he apparently did get $2,000 for it in 1850. A large sum by all accounts. It is possible that he used part of this money to later fight Grover & Baker in court.

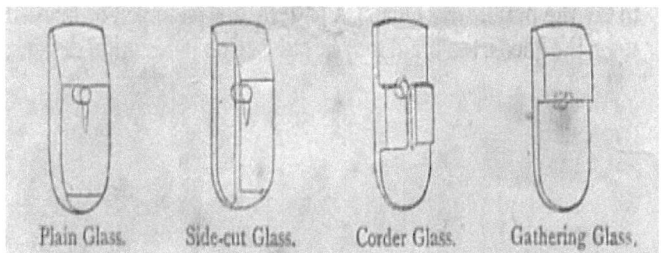

The great idea of glass sewing feet to look through when sewing. It was another 100 years before other competitors made see-through feet.

Here you can see the 1861 patented glass foot and the slightly curving needle that many Wheeler & Wilson models used. The inner glass slipper was replaceable when it was damaged. Also, different shapes were slipped in for other jobs, like shearing, piping and cording. Amazing technology for 1861 as these were hardened glass. They must have been using diamond cutters in their factory.
Image Mike Anderson, Wolfegang's Collectibles.

Having his fingers burnt in his first patent engagement had made him determined not to lose again. When Wilson came up against Grover & Baker, and the fight for his rights to the four-motion-feed, he was prepared. There would be no backing down and capitulating this time.

Orlando Brunson Potter was part of the opposition, working for Grover & Baker's legal department. He was a formidable opponent, but in the strange way that fate works he eventually became close friends with Nathaniel Wheeler, Wilson's future partner. Image Sarony Harpers Magazine.

After a ferocious scuffle lasting several years against Grover & Baker's legal team, headed by their future president, Orlando B. Potter, he eventually won.

This protected his four-motion-feed against any patent infringements for years and stamped his name in history. Well, sewing machine history anyway!

Interestingly, Orlando Brunson Potter and Allen B Wilson became lifelong friends and legend tells that Orlando was so upset at Allen's death he died just a few days later on the 2nd of January 1894. He was the richest man in New York at the time.

The Wheeler & Wilson sewing machine. The silver plated beauty of 1871 was similar to Wilson's 1854 patent. Known as the box base No1W1. You can just see the curved needle.

Here you can see the silver plated arms and superb hand decoration that many of the Wheeler & Wilson machines had. This is the wide-body 1870 model. Image Mike Anderson, Wolfegang's Collectibles.

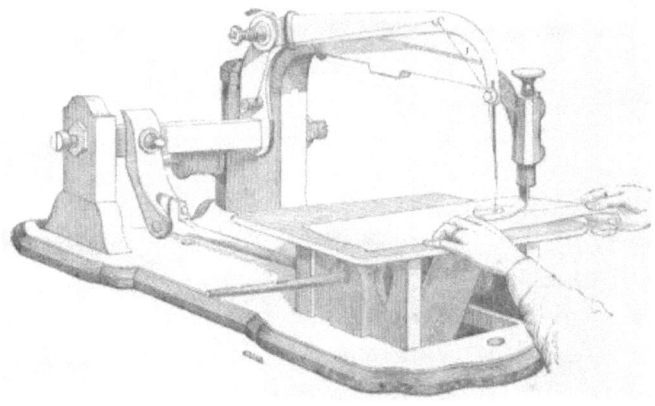

The Wheeler & Wilson Sewing Machine woodcut model No1. Circa 1856. Notice how you sewed on the side, unlike today. Note the long bar protruding on the left of the image. That was to place a grinding stone to reshape the end of the needle when blunt.

Chapter Four

Fate always plays a hand in history. Wilson's journey as one of the great sewing machine pioneers may have been a very different story if he had never met Nathaniel Wheeler. They eventually formed one the most successful sewing machine companies of the 19th Century.

Nathaniel Wheeler

On that business trip to New York in 1850 Nathaniel Wheeler, a keen investor in new ideas, was introduced to the 26 year old inventor, Wilson. As fate had it, via his horrible old partners in the double pointed shuttle, Kline & Lee.

E. E. Lee & Co
(Kline & Lee)
128 Fulton Street
New York

Nathaniel Wheeler was born in Watertown Connecticut on 7th September 1820. His father was a carriage maker and his distant grandfather, Moses Wheeler, had emigrated from England.

Nathaniel also had an inventive mind. He got fed up with the carriage business so he had a go at farming and then manufacturing. In 1849 he became an investor in the Warren & Woodruff Co in Watertown, the company then becoming the Warren, Wheeler & Woodruff Company. Some say he was a silent partner and some say he worked there as manager.

Nathaniel Wheeler was always looking for new ideas to expand into. Interestingly at one point the three partners all lived near each other on the junction of Woodbury Road and The Green in Watertown. Apparently all three houses still exist today!

A lucky meeting in 1850 changed the lives of two men. On a business trip to New York, Allen Benjamin Wilson met Nathaniel Wheeler at Fulton Street and they got on like a house on fire. Wheeler was fascinated with Wilson's 1849 patented 'sewing or seaming lathe'. When it came to light about each other's partner troubles, a cunning plan was hatched.

Wheeler placed an order for 500 machines with Wilson, Kline & Lee. However on the quiet Wheeler & Wilson formed the idea of a new business, picking the best partners from each of the two companies. This was proper 19th Century industrial espionage. Complicated stuff for sure, but many large companies seem to have had these difficult starts.

All Watertown machines were marked A B. Wilson, Watertown, Connecticut.

For a short period, as the first sewing machines were assembled, Wheeler and Wilson were working with George P Woodruff, Alanson Warren, Kline & Lee. The first of Wilson's sewing machines were made but boy was that all about to explode! Two businesses, at least five partners and Wheeler and Wilson the only ones with any the new ideas!

When Wilson later showed Wheeler his latest project for a new rotary hook system with a stationary bobbin, Wheeler saw its immediate potential. Wheeler got Wilson to spend all his efforts quietly preparing a patent model. This was to be the seed for their new partnership when they made their big break.

Just to be sure no one had any idea what they were up to, Wilson kept up a bluff with his old traditional shuttle, but later that year, in November of 1850, he patented an improved double-pointed shuttle, which allowed the shuttle to pick up the needle stitch on its forward and backward movements with elasticity!

This was a clever move. Nobody, except Nathaniel Wheeler, had any idea that his next invention would shake the sewing world. Wilson never followed up with his elastic stitch shuttle as he was hoping that his new rotary hook would take off. It did, like a prairie fire.

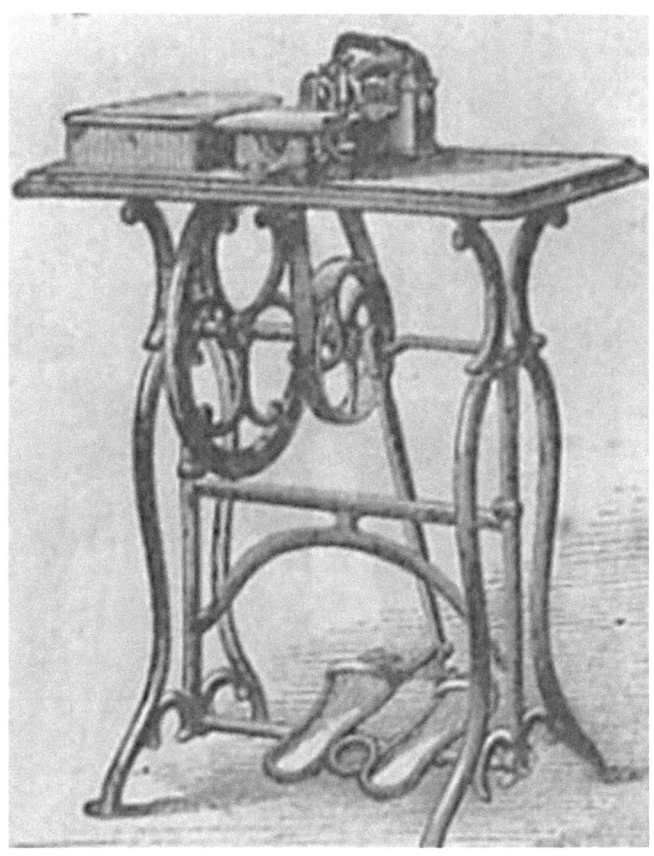

The Wheeler & Wilson Sewing Machine models No1, 2 & 3 were all similar in looks with different finishes, cases and treadle bases. The 'four motion feed' was also added later. The price varied from the top, silver plated model to plain black. This is the most basic model with a slab top.

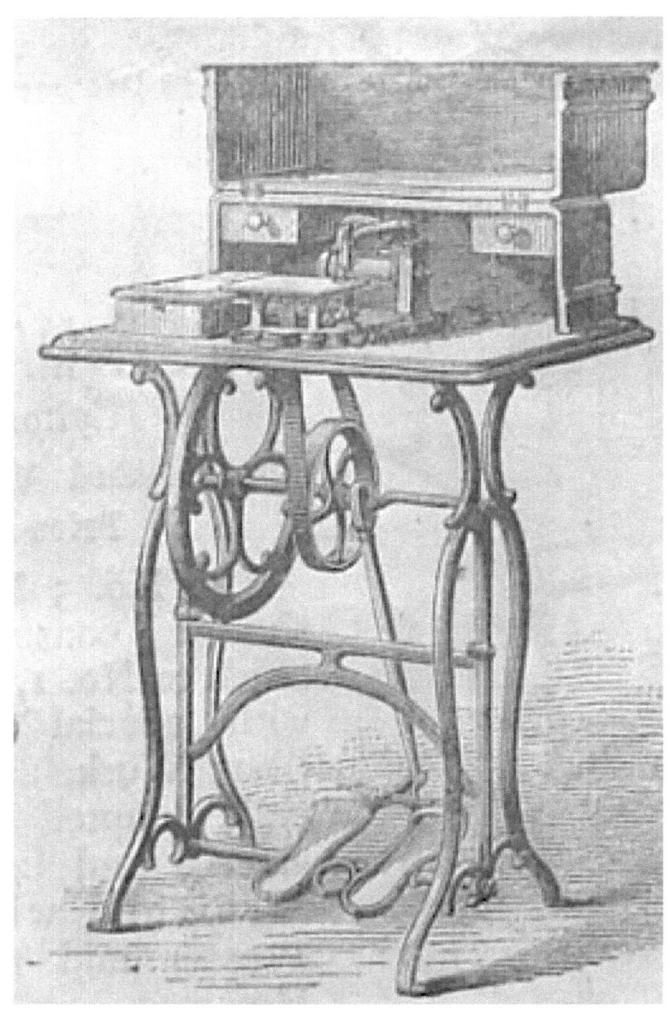

The same machine with a more expensive wooden case and drawers. Note the wide flat belt that was used on Wheeler & Wilson models. There were a few hand machines made using the same flat belting in a small box but I have never seen a W&W one. The only one I came across was the Royal made in England.

This is the Royal sewing machine made in Small Heath, Birmingham, England. It is a direct copy of the Wheeler & Wilson sewing machine. All makers copied the best-selling machines of the day and the W&W was the No1 biggest seller of the period. This is a very unusual hand crank model with a belt drive. Really the machine was better suited for treadle operation. The Royal was far too expensive to sell well in Britain and very few survive. Sewalot Collection.

On the 12th of August 1851 Wilson was granted his improved patent. The beginnings of a great company was forming.

Both Wheeler & Wilson strung their old partners along offering them titbits to sell but by August of 1851, agents across America were quietly being sought for Wilson's new and unique sewing machine.

Nathaniel Wheeler was happy to get the best men behind Wilson and to form a new business to produce quality sewing machines for the masses.

We have to stop for breath for a moment and look at the situation. As demand for these new-fangled gadgets rocketed, more and more patents were being infringed.

The 1850's saw the largest litigation battles in American history. All the original patent holders, known to the media as the 'Sewing Machine Kings', battled it out in court.

The 1850's were the most explosive time in sewing machine history. Giants were rising from the crowd. Men of huge character, like Isaac Singer and Elias

Howe attacked all new sewing machine makers.

Isaac Singer was a formidable opponent in court. He had given half his company away to Edward Clark to fight his battles for him. He hated to lose. Isaac went on to become one of the richest men of his era spending money on wives and mistresses, palaces and racehorses. When he died he was married to a half-French actress 30 years younger than him!

Amongst all this cut-throat dealing and litigation poor Allen B Wilson was trying to kick start his own sewing machine industry!

It was only with the support of the powerhouse of Nathaniel Wheeler that his dream could come true. What Allen B Wilson had in amazing ideas, Nathaniel

Wheeler had in enthusiasm and energy. For a few short years the two men worked hand in hand to build a business that would last beyond both their lives.

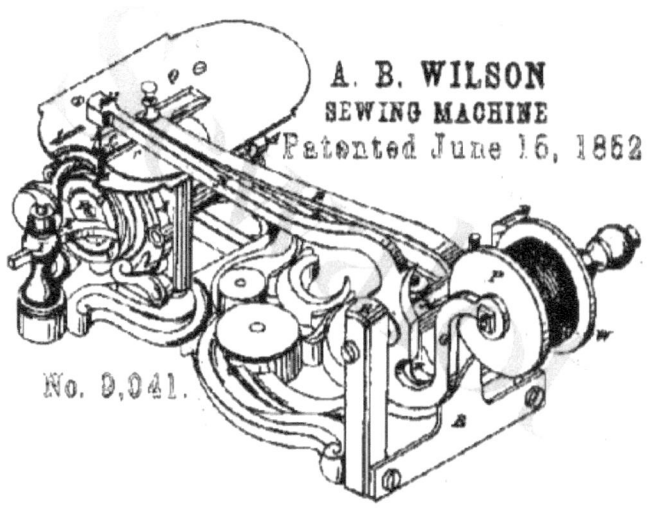

Patent 9,041 by A B Wilson. June 16, 1852.
The Improved Patent Model, Wheeler & Wilson No2, ran from 1852 until the feed was replaced in 1854 (with the Model No3, which ran up until the 1860's and the arrival of the Model No4).

Warren, Wheeler & Wilson Co Watertown. Connecticut

The first sewing machine company seems to have been formed in 1851. It was a partnership between Warren, Woodruff and Wheeler with Wilson soon joining. They worked out of their small factory in Watertown, Connecticut. In 1851 the company had become the

Wheeler, Wilson & Company, then in 1853 it all changed again.

5 October 1853
Wheeler, Wilson & Co becomes the Wheeler & Wilson Manufacturing Company with a capital stock of $160,000.

The officers were, Allen Benjamin Wilson, inventor. Alanson Warren, President, (Wheeler succeeded Warren in 1855). George P Woodruff, Secretary and treasurer. Nathaniel Wheeler, General Manager.

Incidentally, Wilson's amazing four-motion-feed would not be fully secure with patent protection until the end of 1854, when he was finally victorious in court against Grover & Baker. These legal dealings terribly weakened the already frail Wilson. Much like Elias Howe, he suffered terribly from stress and this brought on constant ill health.

A year before, in 1853, Allen B Wilson decided it was just all too much and made plans. He stayed on to fight the court cases and secure his patents for the new company, but his heart was no longer in the sewing machine business.

Chapter Five

Early Retirement

Once again we must step back, this time to 1853, for 1853 was when Allen Benjamin Wilson decided to retire from active participation in the sewing machine industry.

Ill health had forced Allen B Wilson out of the business, however because so many patents were in his name and most of the advertising had been as Wheeler & Wilson (or previously as Wilson) the company had decided to continue with his name on their sewing machines.

From 1853 the Wheeler & Wilson Manufacturing Company was in name only. Wilson had moved on.

Though Wilson would no longer be directly involved with the running of the business he never lost his inventive mind. He also received a salary and patent royalties (now that's what I call early retirement).

Still a young man, Allen B Wilson had started to look into property and land. Maybe he saw the writing on the wall and had seen how well Isaac Singer had done spending his money.

Within ten years of removing himself from the sewing machine field, Wilson was a well-respected land owning property developer.

1870 was the peak for Allen Benjamin Wilson, whose estate was now considerable. Aged 47 he had a personal estate of $130,000 plus another $150,000 in land. Some say this equates to over $100,000,000 today.

History judges Wilson poorly in the context of the huge wealth amassed by other pioneers, like Howe and Singer, but the lad did brilliantly. In 12 short years he changed the way the world sewed and made enough money to retire. Impressive by any standards.

Allen B Wilson built Wilson Hall, completed in 1866. It was a four-storey hotel in North Adams Massachusetts, which had a department store, pharmacy, theatre, lawyers and tailors department. Unfortunately Wilson Hall was destroyed by fire in 1912.

Wilson Hall, North Adams, Massachusetts 1865.
You can see here the hotel nearly finished and Allen B
Wilson's own specially designed central heating pipes
ready for fitting in Wilson Hall.

The Wilson home on the Naugatuck near Waterbury,
Connecticut was later purchased by the city and used as a
community hospital. Note the hat of the photographer by
the tree. Both Wilson and his wife were keen early
amateur photographers.

Allen B Wilson also had a beautiful home just outside Waterbury, Connecticut entered by the long drive, passing a pretty gate lodge. The house had all the latest inventions including his own designed central heating. The back of the house had sweeping views down through the orchards to the Naugatuck River and the town of Waterbury.

1880 daughter Annah (Bennette) with mum Harriet Emeline Brooks (Wilson) and youngest daughter Harriet Ethel Wilson

Wilson's wife Harriet and daughter Annah with a relative Sarah Brooks on their porch circa 1865.

Nathaniel Wheeler also built some superb homes getting larger as his wealth amassed. I've been told that this was one of his holiday homes. His main home was a huge Gothic style home in Golden Hill Bridgeport.

Another super rare image of Allen and wife Harriet on the right and possibly their children to the left. They are at the Naugatuck River near Waterbury. This could have been taken as their house was being built.

Chapter Six

From 1853, Allen B Wilson was more on paper than in the sewing machine business, although he did attend court hearings. Nathaniel Wheeler took up the position of General Manager but in 1855 he rose to the position of President of the Wheeler & Wilson Manufacturing Company (after Warren resigned). A position which Wheeler held until his death.

Wheeler had married twice and had six children. He died at his beautiful Gothic style home in Golden Hill Bridgeport on 31 December 1893.

After his death his son Samuel Wheeler took over as president of the company. The vice presidents were George M. Eames and Isaac Holden. Newton H. Hoyt and Frederick Hurd were treasurers and company secretaries, while William H. Perry was general superintendent.

Chapter Seven

All the main action and excitement was back in the 1850s when a war was raging between the main patent holders. A clever, but very illegal solution was waiting around the corner!

By the middle of the 1850s most of the Wheeler & Wilson patents were secure.

However, the company still had to make sure that Elias Howe (the main litigator of new sewing machine companies) was happy before proceeding. They had temporarily fallen out with Howe after Isaac Singer had talked them into resisting him.

However paying Elias for his patents (like the shuttle and the eye pointed needle) was a small price to pay to allow the growth of their business.

Elias Howe was an extraordinary figure in the sewing world. In his lifetime he only made a handful of machines but became one of the richest men of his age, mainly through legal action. All his wealth could not buy him good health. He was just 48 when he died.

Once all the litigation was sorted between the 'Sewing Machine Kings', the former antagonists formed the infamous Sewing Machine Combination or Cartel. Elias Howe, Isaac Singer, Grover & Baker, Wheeler & Wilson all got together in Albany and came up with a cunning plan.

Instead of the big boys spending all their time and money in court attacking each other, they would get together and attack all the other fledgling sewing machine manufacturers. The Albany Agreement became notorious.

This illegal combination was eventually destroyed, but for many years this monopoly effectively crippled nearly all new competition. Some makers, like Charles Raymond, escaped from America altogether. He set up his very successful business in Canada, away from the shenanigans going on over the border.

On the left is an 1850's model and on the right an 1872 Wheeler & Wilson. It is difficult from the advertising to see the difference in size but in real life it is easy. The early heads were far larger. Image Mike Anderson, Wolfegang's Collectibles.

Chapter Eight

Let's ignore the legal proceedings for a moment and step back once again to the early 1850s, so that we can look at what machines Wheeler & Wilson were making. Things were heating up with the development of the sewing machine. At this time Allen B. Wilson was working hard in his fledgling business.

In 1852 the first of 200 Wilson sewing machines were made, even though the patent had not been issued for the stationary bobbin until June 15th of that year. With Howe's fees, each new machine sold for $125, a year's wage! Boy that was expensive. Each machine cost the price of a new car today.

Their first machine, the No1, grew into 11 different styles and cabinets and looked suspiciously similar to the Grover & Baker machines, even down to the needle patented by Elias Howe in his weird Indian dream. You will have to read his story to find out how the poor farmer's son rose to one of the wealthiest men in America, all supposedly from a dream!

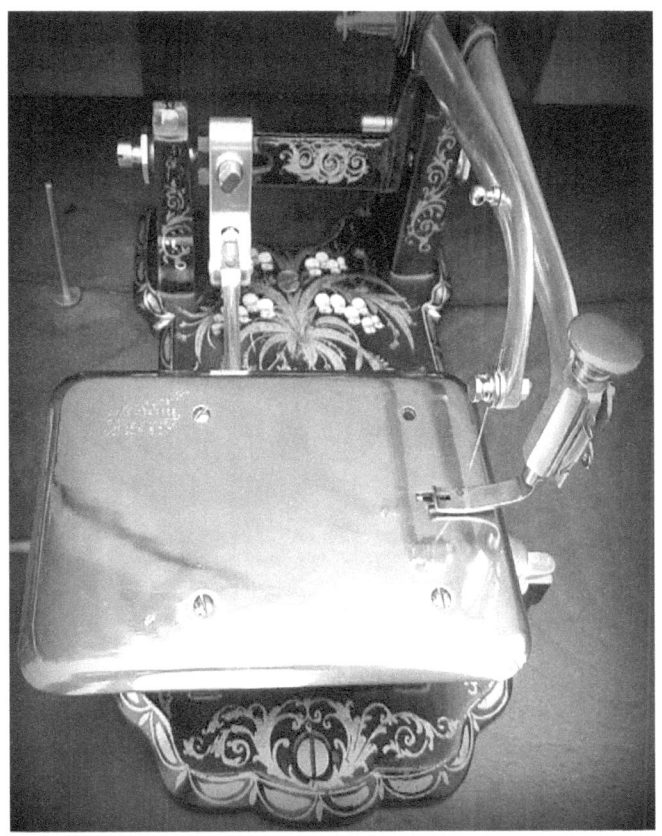

This stunning Wheeler & Wilson is one of the finest I have ever seen. It is decorated in the 1859 fern pattern. You can see how this beautiful machine would appeal to the customers. It is pure opulence with a price tag to match. Notice how the operator sits to the side of the machine. I imagine that you would soon get used to the idea of the work moving from left-to-right. Image courtesy of Mike Anderson, Wolfegang's Collectibles.

The Wheeler & Wilson No3 model was a triumph. It contained all the previous inventions and patents as well as the phenomenal 'four motion feed' (that changed the sewing world). Although the diagram shows a straight needle, models 1, 2 & 3 were all using a slightly curved needle. I love the instructions, "If the needle does not pass through the centre of the hole in the bed, bend it until it does!"

Patent 12116
19th December 1854
Four-motion feed for a sewing machine
Granted to
A. B. Wilson

The Model No4 had come along by the 1860's and was a great success, with a larger bed and less thread snagging. Priced at $75 it was only $10 more than the domestic. It ran almost unaltered from about 1861 to 1877. When the patents expired it was widely copied.

The Wheeler & Wilson Model 4 heavy duty. $75 in 1861 with a later W&W German copy below. Notice the subtle difference of the needlebar assembly for the different models.

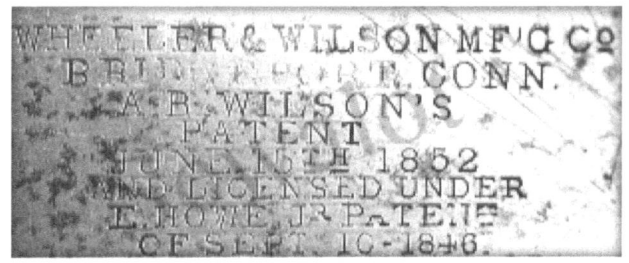

The early Wheeler & Wilson sewing machines incorporated the Howe 1846 patent. No sewing machine could be made without Elias's needle patent, which meant he became rich without actually making sewing machines.

Wilson oversaw each machine to make sure they were perfect and by the end of 1852 around 500 machines a year were being built. Wilson was working too hard and too long and the pressure on his body was telling.

Every machine was hand built and perfected before leaving the factory. This was known as file-to-fit. Better machines would see the rise of mass production, eliminating the need to make each part fit a particular machine. With mass production any part from any machine could fit perfectly into any other.

Wheeler & Wilson sewing machine model 5 long-arm treadle for specialised heavy sewing. $85 in 1860. The model number are more confusing now as they mixed domestic models with industrials.

Wheeler & Wilson sewing machine N06, heavy flatbed circa 1870

Model 8 Wheeler & Wilson sewing machines on a steam driven shaft assembly for factory use, circa 1888. The steam boilers would have been housed in another building close to coal and water supplies. The power was then transferred through shafts into the sewing rooms. The boilers would be kept up to 'steam' almost 24 hours a day so that the workers could sit in shifts. Mass production of clothing had arrived.

New England Office
1872
228 Washington Street,
Boston.
H C Hayden, Agent.

Chapter Nine

Bridgeport

Business was booming and sales soared. One of the company's bigger customers was Oliver Winchester of the Winchester repeating rifle fame. Oliver Winchester had made much of his fortune hand producing shirts in huge hand sewing factories, but once he saw the benefits of the Wheeler & Wilson sewing machine, he invested heavily in them. With sewing machines, instead of a day, a shirt could be made in less than an hour at one of his factories! The prices consequently dropped and business boomed.

In 1856 new premises were needed for Wheeler & Wilson, to deal with the ever expanding order book. The company slowly moved its production down the road to Bridgeport, which had better transport links and a railway.

The former clockmakers of Jerome & Co had the perfect factory, which Wheeler & Wilson snapped up and enlarged, year after year until it looked like a small town, employing thousands of highly skilled workers. Cleverly, Wheeler employed the skills of J. D. Alvord from the Sharps Rifle Works. He oversaw the design of many of the new machines to perfect mass production.

By 1858/59 over 18,000 machines had been produced with various new models appearing in all shapes and sizes, from lightweight machines for silk to heavy industrial machines for horse blankets and leather.

The company's capital stock was increased in July of 1859 to $400,000. A superb new showroom opened in New York at 343 Broadway and the company was on its way to becoming the largest sewing machine maker in the world.

Interestingly as the company boomed it moved continually along Broadway in New York.

1857, 343 Broadway, NY.
1858, 345 Broadway, NY.
1859-64, 505 Broadway, NY.
1864-1872, 625 Broadway, NY.
1872-1874, 635 Broadway, NY.
1874-1886, 44 East 14th St, Union Square, NY.
1886, 833 Broadway, NY.

Wheeler & Wilson sewing machine model 6 cylinder arm circa 1870. Perfect for harness and leather work.

Note how the operator can slip right into the machine and become almost part of it. It was brilliant design like this that kept the company in front of its competition. Amazingly some of these machines still survive and are in regular use. You can bypass the treadle too. New electronic servo motors allow the operator to choose the speed they need for the job they are doing and drop onto these old beauties perfectly.

Chapter Ten

Civil War

THE SEWING-MACHINE.

"A most wonderful invention indeed, ladies! The sewing machine executes the work so efficiently that upon my word I think there will be nothing left for women to do but improve their intellect." Punch 1866. I don't think he would last long saying that today!

It is an interesting point to make that during the American Civil War sewing machine sales boomed for some manufacturers.

For example in 1863 Wheeler & Wilson produced over 30,000 sewing machines and each following year the business expanded at a massive rate. It was the start of a glorious few years for the Wheeler & Wilson Company.

1865, Over 50,000 sewing machines sold by Wheeler & Wilson in 12 months.

50,000 sewing machines made and sold in 12 months! Not bad going for the end of the Civil War. Like Willcox & Gibbs, sales had boomed during the conflict. It is a side to the war that we never really hear about.

During the boom years the Wheeler & Wilson Company was perfecting its mass production techniques. This meant it could make machines better, faster, and cheaper!

In June of 1864 a special charter was granted by the State of Connecticut, allowing Wheeler & Wilson to increase its stock to $1.000.000.

In 1865 the Wheeler & Wilson Company licenced for sale 39,157 sewing machines. In that same year Singers only managed 26,340. But that was about to change.

In 1865 the Singer Company had brought out the fabulous Singer 12, which was an instant best-seller. From now on Singers would be nipping at their heels.

One thing that is worth pointing out is that in 1866 the Wheeler & Wilson prices were still extraordinary.

The most expensive machine in their range, the silver plated model No1, in a walnut and rosewood cabinet was almost $200. Well over two years' wages! I mean that's the price of a new car today!

Some of the cabinets that the Wheeler & Wilson sewing machines came in are of unrivalled quality, looking more like Chippendale furniture. They had some very highly skilled craftsmen at their factory, just on the woodworking side.

In the 1890's at the huge Singer factory in Scotland over 2,000 craftsmen worked on the wood, cabinet and cases side.

The Wheeler & Wilson sewing machines were beautiful and practical and led the world of sewing machines with elegance and style. Note how you sew sideways on the Wheeler & Wilson from left to right. This machine is in a stunning purpose built cabinet made to resemble writing desk for the parlour.

SCHEDULE OF PRICES: 1866

No. 1 Machine, Silver Plated, with

Full Case, Polished Rosewood, narrow	$110 00
Full Case, Polished Rosewood, and Drawers	135 00
Full Case, Polished Rosewood, and Drawers, lined	195 00
Full Case, Polished Black Walnut or Mahogany	105 00
Full Case, Polished Black Walnut or Mahogany, and Drawers	110 00
Full Case, Polished Black Walnut or Mahogany, Concealed Hinges	115 00
Full Case, Polished Black Walnut or Mahogany, Concealed Hinges and Drawers	125 00

Can you imagine paying these sums today! The Wheeler & Wilson in 1866 cost the price of a new car! You could get cheaper models in plain black from $65, still a small fortune back then.

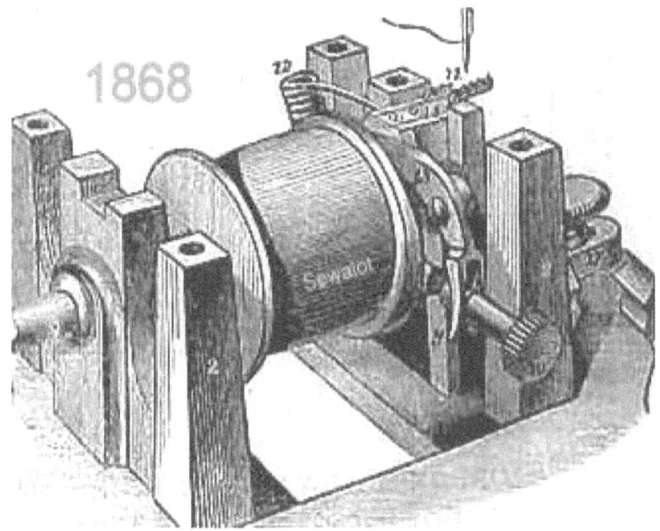

This advert from 1868 is showing a method of adjusting the feed length. The knurled knob adjusted the height and length of the feed. Notice the huge wooden 'bobbin' cylinder for the flat belt to turn the machine.

The best year for Wheeler & Wilson was probably 1875 with the all-new model 8. That year, nearly 300,000 machines were produced in a 12 month period. No other manufacturing business in history had ever completed such an amazing feat.

The Wheeler & Wilson Company became the largest manufacturer of sewing machines in the world. Mind you, Singer were still close behind and expanding globally at a phenomenal rate. Because Singer built sewing machines in many countries they became far more recession proof. Consequently they survived, even thrived. Many businesses manufacturing in just one country suffered when that country had a recession.

Over 12 models and dozens of options of Wheeler & Wilson sewing machines were now available. They became one of the best money could buy.

Wheeler & Wilson sewing machine N06 D6W, heavy flatbed circa 1870, you can see the start of the model 8, which was only a few years away but it is far removed from the similarly named No6 above and a real Wheeler & Wilson Model 6 below.

The real deal, a tough flatbed Wheeler & Wilson Model D6 capable of handling a large variety of work. Interestingly they never called their large machines industrials. The term 'industrial sewing machine' had not yet been invented. Instead they would describe the sort of work that the machine was capable of handling. Wheeler & Wilson noted that 'for beauty of stitch and firmness of seam there is no equal.' The company also supplied plans with each machine for steam power!

The Wheeler & Wilson D6 barrel arm machine was for heavy saddle, boot, and strap work. These machines turn up occasionally but they are rare today. This model would compete directly with the Singer model 29.

In fact every manufacturer from the 1860's onward produced a range of industrial machines. Isaac Singer actually started with an industrial and worked backward towards a domestic machine, perfecting the Singer 12 by 1865.

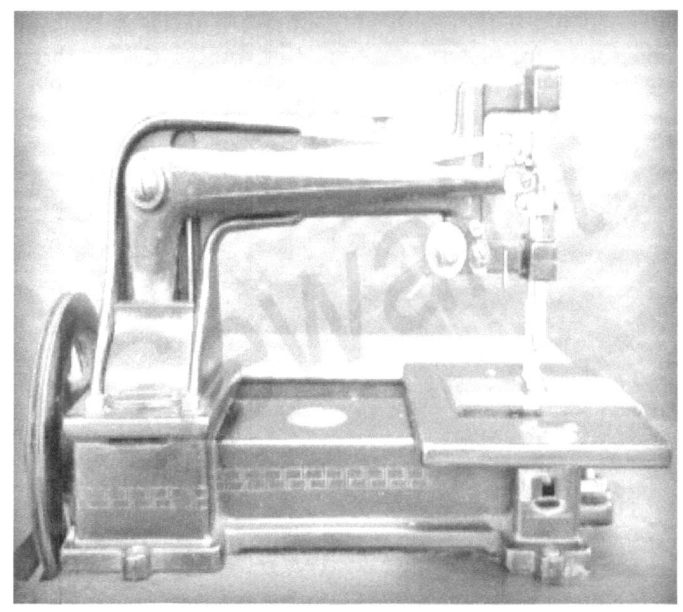

The Wheeler & Wilson model D6 from the rear.

Mike Anderson from Wolfgang's Collectibles came up with this great bit of information: There were 3 distinct model W&W 6 treadles made.

1. The cylinder bed machine

2. The flat bed with fly wheel on the LEFT-HAND treadle leg.

3. The flat bed machine with the flywheel at the back-side of the treadle... similar to a Grover & Baker set-up.

Chapter Eleven

By the late 1870s all the patents that could be, had been extended, and finally exhausted. For the first time this allowed any sewing machine business to compete on a level playing field, without having to pay royalties.

There was a boom in small independent sewing machine makers, undercutting the big companies. True mass production in sewing machines had arrived and every industrial country in the world started producing the 'must have' machine of the age. The sewing machine went on to be crowned one of the most useful inventions of the 19th Century.

The result of mass production from unlimited small sewing machine makers would mean that in a few years production would start to fall dramatically at Wheeler & Wilson.

Though sales would drop, the company never let its quality slip. The Wheeler & Wilson machines were super smooth with some later models even had ball bearings! This was way ahead of the competition. Singer eventually copied the idea and used it on their industrial machines like the model 660A2.

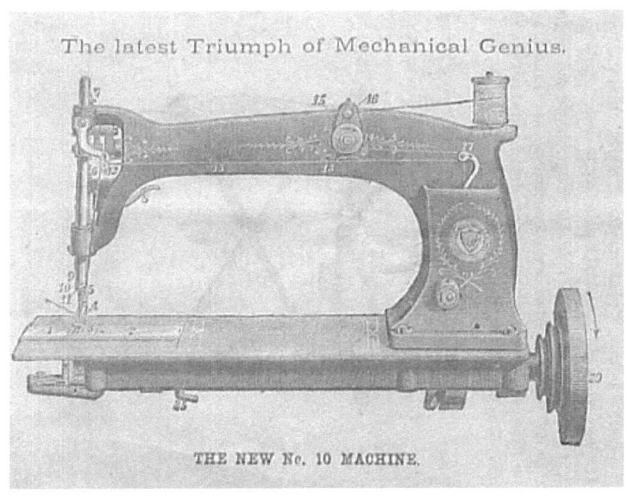

A general purpose tailoring machine, semi industrial. The Wheeler & Wilson sewing machine No 10 a modern miracle in 1886 with a large bed area. Great for quilts and bedding.

A handful of patent models still exist but they are in rare collections. The early model No 1 is my personal weakness. It was made for decades with only minor changes and is still one of the finest looking and best early sewing machines for actually sewing with. These superb pieces of engineering won countless awards.

The curved needles of the early model Wheeler & Wilson machines are a real pain to find today and that could be the end of a fine machine if none can be sourced. Still, as an object of early engineering, and one of the first mass produced mechanical marvels, the Wheeler & Wilson No1 holds its place as a stunning piece of 19th Century engineering.

The amazing Wheeler & Wilson D12 was a zig-zag industrial version of the N12, capable of multiple tasks from sailcloth to quilts. Singer later used many of its unique ideas on their models right up to the Singer 206 in the 1950s

Chapter Twelve

Wheeler & Wilson sewing machines were simply superb and won an amazing amount of medals all over the world. Their production methods had been improving since 1851 and by the 1870's they were making the best quality sewing machines that money could buy. Unfortunately all around the world small factories were pinching their market. Few 'big boys' would survive this monumental shift from large to small manufacturers.

Some Wheeler & Wilson Medals

Paris 1861 & 1867, Gold International Award.

London 1862

Vienna Exhibition 1873, Imperial order of Francis Joseph, Grand medal of Progress and Medal of Merit.

American Institute of New York, Gold medal of Honour.

Maryland Institute Gold and silver Medal.

Georgia State Fair, Nov 1873 silver medal in leather stitching.

British Awards

July 1874 Silver Cup at the United East Lothian Agricultural Society First Prize.

Aug 1874 First Prize Bury Agricultural Show.

Manchester & Liverpool Agricultural Show 1874, Silver Medal W&W No 6 for Excellence in Manufacture.

September 1874 Cheshire Agricultural Show First Prize.

Wheeler & Wilson
Chief European Offices
11-21 Paul Street, Finsbury, London.

Wheeler & Wilson had expanded into Europe early with their first office opening to an eager market in 1859 at 13 Finsbury Place, London. As the company prospered they bought plush offices in Regents Street, London.

The Head Office at that time was in Queen Victoria Street. By the end of the Victorian era they had branches and agents all over Britain and Europe, with their Head Offices along Paul Street in London.

Wheeler & Wilson model No 7 flatbed heavy duty machine for leather work. Circa 1876. Note it is still a treadle. Can you imagine working leather on that! You would end up with legs like an Olympic athlete.

Medals galore for the fabulous W&W machines

Also Wheeler & Wilson had agents, stores, departments and depots all over the world, including 26 in Britain!

The 1861 Paris gold award.

A very special appointment. The Princess of Wales, later Queen Alexandra. She was a keen sewer and supported several manufacturers including her favourite Jones Company in Manchester.

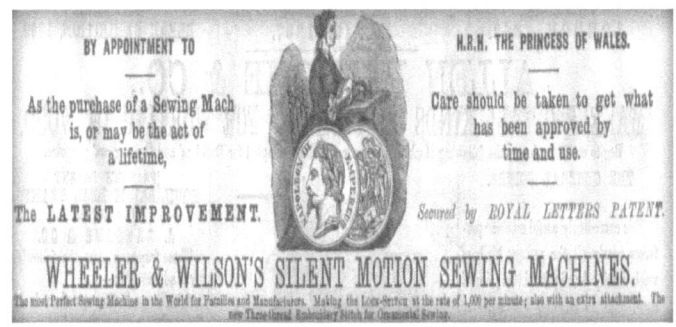

1873 Wheeler & Wilson prices UK Prices from £6 or terms at 2s 6d a week. Boot maker's machines also available on terms from £9.

Chapter Thirteen

The Wheeler & Wilson No 8

The finest machine in the World

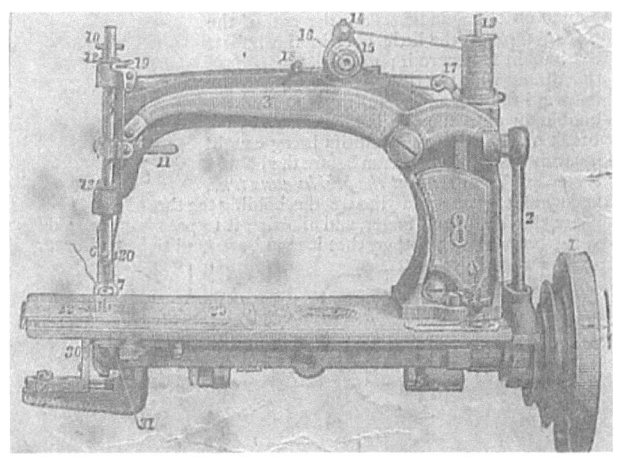

The Wheeler & Wilson sewing machine No 8 circa 1876. This machine should have been a huge success but its industrial looks and huge price led to less sales than expected. It sewed like a dream and even today they are great machines. More patents were granted on this and other models as improvements carried on. March 28th 1876 Patent 175463

The Wheeler & Wilson sewing machine model No8 woodcut circa 1876

By the 1870's most sewing machines being made, by countless factories around the globe, were now reliable. Some were being offered with lifetime guarantees (like the German Pfaff). Sales were now all about price and promotion.

Wheeler & Wilson sewing machine sales dropped in inverse proportion to the new manufacturers starting up. In Germany in 1850 for example, there was one sewing machine maker, by 1900 there were over 300! All this was destroying the traditional big boys.

The only company that bucked the trend was Singer. They invested over a million dollars a year in advertising. They expanded at a phenomenal rate into any emerging country. They brought law suits against smaller firms, destroying competition where they could. They built new factories around the world to save on the cost of moving machines great distances and kept on improving their sewing machines. All-in-all the perfect company.

By the end of the 19th Century Wheeler & Wilson was just a shell of its former self. In 1904 discussions were held between the Singer Company and Wheeler & Wilson. The Singer Company were pleased to grab an opportunity to use the huge Bridgeport factory that Wheeler & Wilson had. They were also delighted to get their hands on the phenomenal patents that the company had been continually getting for improvements to sewing machines.

By buying up Wheeler & Wilson, the Singer Company could carry on expanding and remove its main competitor in one swoop. Perfect business really.

Let's step back a second and look at the decline of Wheeler & Wilson.

Although by the early 1870's W&W were losing ground to their competitors, their temporary saviours were their in-house research and design team, the House Brothers.

James A House and Henry A House basically took the best ideas from the early W&W machines and put them into the all-new all singing and dancing models. In May of 1871 they were awarded the patent for their improved rotary hook sewing machine with an eye-pointed needle.

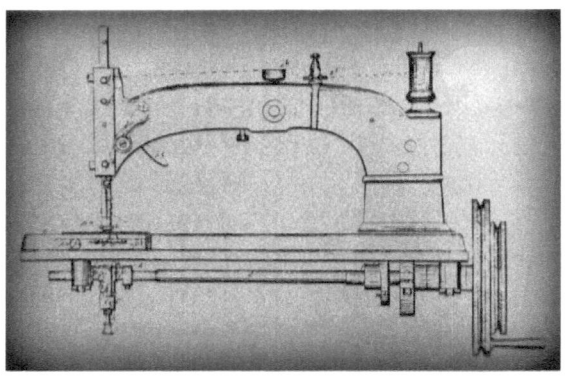

Patent No 114294 was patented in 1871 by J A & H A House, inventors working for Wheeler & Wilson. It was for improvements to the rotary hook that allowed for different threads to be used without snagging. It looks similar to the later Model 10 but it may have been the improvements that were incorporated into the Wheeler & Wilson sewing machines, not the actual 'look' of the patent model above. J A House Patents, April 1st 1870, March 5th 1873, Dec 16th 1873.

Their work culminated with the astonishing Wheeler & Wilson No8, possibly the finest sewing machine on the market at the time. Some of the early needle plates were stamped with J A & H A House's patents.

The Wheeler & Wilson No8 was a great success in the 1870's. As a near perfect sewing machine it put Wheeler & Wilson temporarily back amongst the top players in the field. Its rotary hook, straight needle, solid take-up lever mechanism and positive feed had brought the Wheeler & Wilson sewing machines into the modern world.

Some go as far as to say that the Wheeler & Wilson Model 8 was the finest sewing machine of its time. It could handle most types of work with ease. It looked cool too. One problem, top quality engineering, even with the latest mass production techniques still cost lots of money.

Boston announcement. Due to expansion
The Wheeler & Wilson New England Agency
will be moving from Tremont Street to
594 Washington Street.
(Immediately adjoining The Globe Theatre)

With over a million Wheeler & Wilson sewing machines made, every collector would like to have at least one model in their collection. I know a few collectors that have dozens. If you do come across an early model it may be silver plated or even have Mother of Pearl inlay!

The Wheeler & Wilson No 8 in treadle form.

A rare postcard from 1866 showing a travelling journeyman and his young assistant who repaired and sold Wheeler & Wilson sewing machines. These were often journeymen paid by the day.

After the amazing Wheeler & Wilson Model 8 the company was going to set the sewing world on fire with an even better sewing machine. Some say the finest sewing machine of the age! The Wheeler & Wilson model 9 or D-9.

You are about to find out that I do go on about the Wheeler & Wilson model 9. Anyone who has one will understand. It is fabulous.

Turn drudgery into pastime was an advertising master stroke. By moving into the hobby industry and away from the need to sew, Wheeler & Wilson were hoping to open up a whole new market. You could possibly trace sewing for pleasure on sewing machines back to around this era.

Chapter Fourteen
Model No9
The Perfect Machine

Launched in the late 1880s the Wheeler & Wilson 9 or D-9 was a world-beating trend setter. It took on all opposition and blew them away. Some of the bearings were even frictionless ball bearings! It was probably this model that forced the Singer Company into its takeover in 1905. I'm not sure what the parrot has to do with sewing!

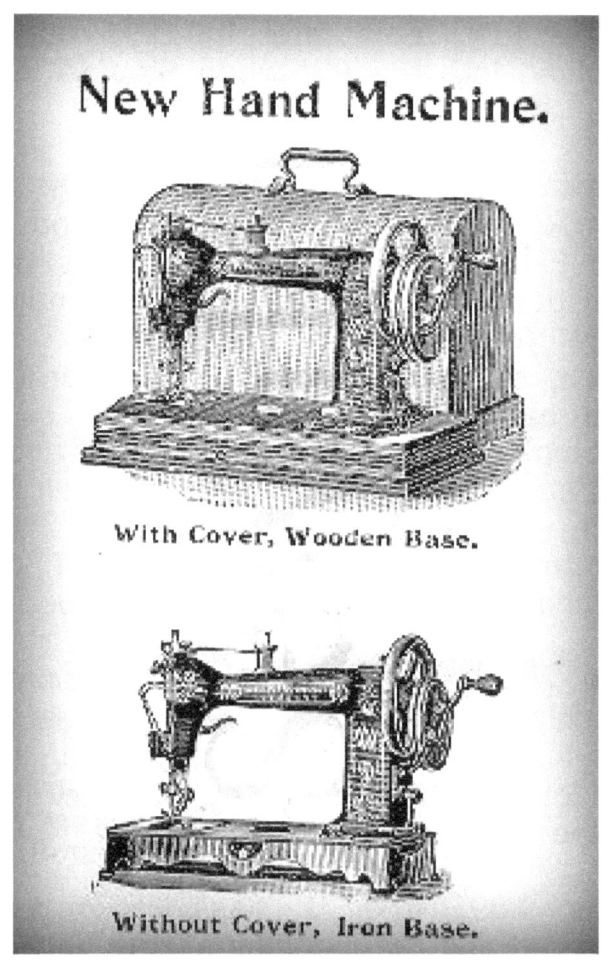

A rare woodcut of the iron base model 9. I have never seen a real one!

Following close on the heels of the model 8 was the astounding model 9.

Let me quickly explain the Wheeler & Wilson model 9, D9 or 9W. The model 9 was first launched in 1887. The D9 came out a few years later in 1895 and it evolved into the 9W when Singers took over after 1905.

Basically the W&W model 9 was to compete with the increasing demand for the European 'high-arm' machines that had been flooding in to America, mainly from Germany. The high-arm of the European models gave more clearance and people loved them. Quilters could roll up an entire quilt and sew it under the arm of the machine. You can tell the early models as they had the stitch length adjuster on the pillar of the machine and the later ones on the bed.

The W&W model D9 was an improved machine with a smaller hand wheel. It used much of the mechanism of the model 11 that had proved so successful in the industrial and factory world.

After 1895 the D classification was dropped and the machine simply became once more the W&W model 9. After Singers takeover the class was marked Singer 9W for several years. More of that on a mo. Now let's look at the beginning of this miracle of 19th Century engineering. The finest machine in the world in 1887.

In 1887 this was the finest sewing machine on the planet. You can see that this is the slightly later model as the stitch length adjuster is on the bed, not the pillar of the machine as in the early models. All the model 9 machines had a brass W&W badge in the bed, even some of the later Singer machines had them as all the old stock was used up after 1905. Note the unusual hand wheel on this model.

Nathaniel Wheeler had been working on improvements with his inventor at the time, Wilbur F Dial, through 1883 to 1886, applying for several patent improvements to the machines. Even after its launch they gained further patents for it. The Model D9 sewing machine improved and improved until it became the most amazing machine.

The Wheeler & Wilson model 9 or D9, later to become the Singer 9W

The Wheeler & Wilson model D-9 sewing machine was a huge success in terms of engineering, though far too expensive to produce competitively. It ran right up until the 1930's, though from 1905 it was dressed in Singer apparel. A rare few model 9's turn up with Singer decals, but still a Wheeler & Wilson brass badge in the bed.

For years people thought these were fakes but they were probably just old stock being used up.

The best sewing machine in the world

Patents
May 1st 1883
Oct 23rd 1883
Dec 2nd 1884
Aug 18th 1885
July 17th 1888
March 25 1890
August 2nd 1892

The D9 was packed with so many state-of-the-art modifications, features and patents that no machine could come close to the perfection it attained in stitching. Singer continued the D9 as the Singer 9W.

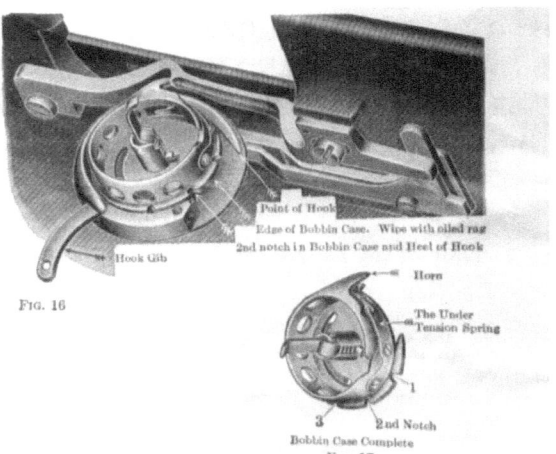

FIG. 16

Besides the amazing bearings the model D-9 had this stupendous hook assembly. When the Singer research and development team saw this they must have stood back in amazement. This system later went onto many Singer machines including larger industrials like the model 211G and domestics like the Singer 201 (but held horizontally to allow for a drop in bobbin). Sewing machine technicians will also notice the similarity between this hook and the Singer Featherweight 221 and 222k. It was the first truly perfect stitch system.

The narrow high-arm on the D-9 allowed full sized quilts to be sewn with ease and it could also handle any fabric from silk to sack-cloth. Once again Singer's competitor had snatched the advantage. By 1892 the multi-award winning Wheeler & Wilson model No9 cost around $60, but with some new cheap sewing machines retailing for just $2 this was far too much! The price of the model 9 was the company's downfall and added one more nail in the coffin of the giant firm.

The Wheeler & Wilson No 9 advert 1902. A stitch in time saves nine! My mum tried this on me a few times!

A Wheeler & Wilson agency circa 1880s and the lower one a Singer shop selling Wheeler & Wilson machines in 1906.

When the Singer Company research and development team first laid eyes on the Wheeler & Wilson D9 they must have had palpitations! Almost every idea was protected by new patents.

It was only by the takeover of the W&W business that the Singer Company managed to get their hands on the model No9. If you could afford one of these beauties it was bullet proof. A 100 year old Wheeler & Wilson will still sew like new. The price may explain why so few turn up today, especially in hand form.

Wheeler & Wilson machines could also be bought on the new-fangled part-payment scheme invented by Isaac Singer's partner Edward Clark.

When Singer machines were not selling (because of the price) Clark came up with the idea of giving the machines away and letting the customer pay for them at their convenience, over one, five, 10, 15 and 20 years! This layaway or hire purchase scheme was brand new in the 1850's.

It transformed sales almost overnight. I met a woman who paid for her sewing machine from 1921, until her last payment in 1941! Astounding but true.

I love this sewing machine cabinet that was made to fit into a library complete with fake books. Does a real one survive? At a cost of a year's average wage there may be one tucked away somewhere in a grand old house just waiting to be discovered.

The Wheeler & Wilson cabinets were true works of craftsmanship. They say over 250 highly skilled craftsmen worked on the wooden parts of the sewing machines at the factory. This cabinet would sit perfectly in a study or drawing room and then transform into a working machine.

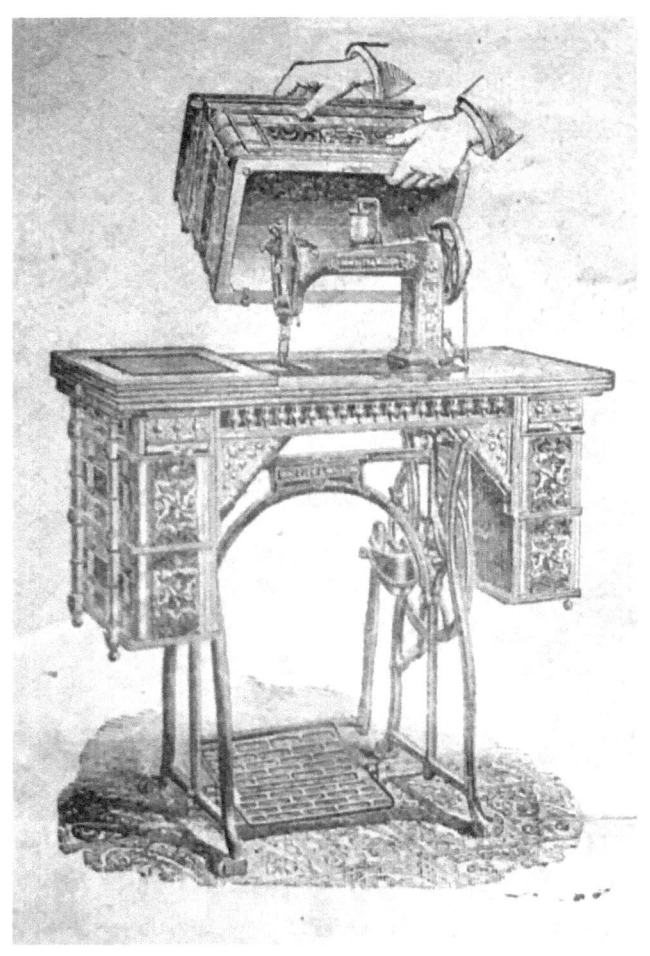

Of course the cheapest W & W No9 was the basic cast iron open treadle. They were still a masterpiece of design and elegance but were obviously visible as a sewing machine even with the lid on.

A sweet advert for a toy Wheeler & Wilson machine. I have never seen one. Do any exist today? If so they would be worth a king's ransom.

By 1905, contracts were exchanged and Singer took over the huge Wheeler & Wilson factory. They carried on with the No 9 along with several other models including industrials. Singer also used the Bridgeport factories to make parts for their own machines.

The No 9 continued to sell in reasonable numbers dressed in Singer clothes up until the outbreak of World War Two. I have a Singer/Wheeler & Wilson in my Sewalot Collection. It has all the Singer marking and decals but still the Wheeler & Wilson brass badge in the bed. More importantly some of the Wheeler & Wilson sewing machine parts were used on Singer models.

There is some dispute amongst collectors exactly how long the enormous factory at Bridgeport, Connecticut did stay open. Some say the sprawling fifteen and a half acre site started to be downsized and be demolished after the First World War, some say during the 1930s depression era. Others say that part of it carried on for decades making Singer parts and heavy machines. Recent documents indicate it may have still been in use throughout the 1960s and declined as the Singer Company did.

At its height the Wheeler & Wilson Company had at least 15 models in their range (as well as many model derivatives) including a fabulous fully automatic buttonhole machine for factory use. At this time the Bridgeport factories were taking up the space of 16 football fields. A sight to behold! Tours of the factory were held twice fortnightly.

However as demand reduced so did the range.

By 1894 Wheeler & Wilson had cut right back on models 5, 7, 8 and 10. The glass foot on the models was removed and replaced with cheaper steel feet.

Wheeler & Wilson even made a short bed model D9 the 9H4, which are quite rare today, almost identical to the D9 except about two thirds the size.

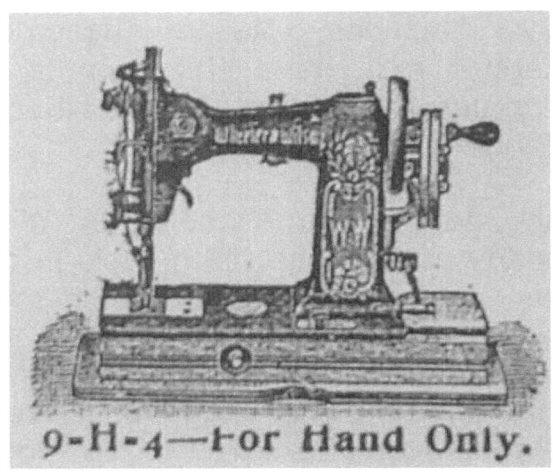

The half-size half-pint Wheeler & Wilson 9H sewing machine is a rare model today. All the benefits of a full size machine but half the weight.

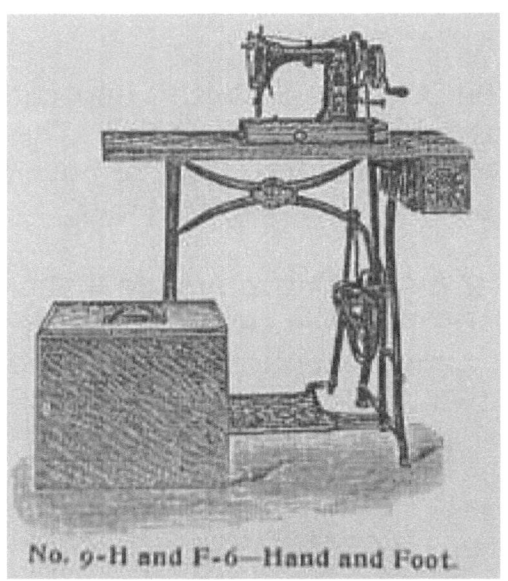

Lintner & Sporborg were general agents for Wheeler & Wilson in Gloversville, NY. They sold the popular W&W No9, supplied parts and servicing.

The perfect scene. A ladies parlour with a Wheeler & Wilson No 9 waiting for alterations. How the other half live eh! Oh to have staff...

Chapter Fifteen

Models & Serial Numbers

Wheeler & Wilson Manufacturing Serial Numbers from 1851 (Smithsonian). These serial numbers are for the original curved needle models 1-5.

1851 1-200

1852 201-650

1853 651-1449

1854 1450-2205

1855 2206-3376

1856 3377-5586

1857 5587-10177

1858 10178-18155

1859 18156-39461

1860 39462-64563

1861 64564-83119

1862   83120-111321

1863   111322-141099

1864   141100-181161

1865   181162-220318

It is interesting to note that during the Civil War Wheeler & Wilson made over 180,000 sewing machines. It shows that sales were on a boom during this difficult period.

1866   220319-270450

1867   270451-308505

1868   308506-357856

1869   357857-436722

1870   436723-519930

1871   519931-648456

1872   648457-822545

1873   822546-941735

1874   941736-1034563. Over 100,000 sewing machines made by the company in 12 months!

1875   1034564-1318303

1876   1138304-1247300

To make matters far more confusing, Wheeler & Wilson started using letters as well as numbers for their models, so you now have model 6 and model D6, its industrial counterpart. Hence the numbers altered once more. This has caused a lot of frustration in dating Wheeler & Wilson models from around 1876 until 1905. When Singer started putting their letters on Wheeler & Wilson models it became even more confusing!

It is also interesting to note that while any company is expanding they are great at showing production numbers. Banks can see the expansion and are happy to lend and invest.

However when the tide turns and production numbers start to fall it is amazing how many companies start to lose their records! A poor production year and a drop in numbers makes shareholders nervous. Consequently records suddenly get lost and my job gets a whole lot harder!

My friend Wayne Schmidt did some brilliant research on the W&W model 9 serial numbers. Now, due to unknown exact start date of the W&W model 9 production run and the exact date they finished, this list can only be taken as a guide. Also did they start at 0? However that said this is the best guide so far.

The serial numbers for the first W&W model 9 that ran from 1887-1894 are roughly as follows.

1-30,000 - 1887

30,001-68,000 - 1888

68,001-102,000 - 1889

102,001-140,000 - 1890

140,001-174,000 - 1891

174,001-208,000 - 1892

208,001-242,000 - 1893

242,001-270,000 – 1894

Wheeler & Wilson model D9 serial numbers, 1895-1904.

2,270,001-2,355,000 - 1895

2,355,001-2,440,000 - 1896

2,440,001-2,525,000 - 1897

2,525,001-2,610,000 - 1898

2,610,001-2,695,000 - 1899

2,695,001-2,780,000 - 1900

2,780,001-2,865,000 - 1901

2,865,001-2,950,000 - 1902

2,950,001-3,120,000 - 1903

3,120,000-3,205,000 – 1904

I have no data on the post 1905 Singer 9W serial numbers from Bridgeport. They could be incorporated with the standard Singer dating numbers.

Sewing Machine Models

This is not a complete list but close

1850, Wilson Patent model 1, The Sewing Lathe.
1852, Improved Sewing Lathe 2, patent 9041.
1853, Patent Sewing Lathe 3.
1854, Improved sewing lathe 4, patent 12116.
1855-1893, No1.
1855-1887, No2.
1855-1865, No3.
1858, Elliptic sewing machine. Similar to the George Washington Gates model made in Toronto, Canada.
1860-1876, No4.
1860-1878, No5.

1872-1883, No6.
1878-1883, No7.
1876-1890, No8.
1887-1895, No9.
1895-1905, NoD9 and later Singer 9W.
1880-1885, No10.
1886-1890, D10.
1892-1905, No11 and later Singer W11.
1887-1891, No12 & N12, A light industrial version of the No9 with a barrel arm as well if required.
1890-1905, D12 and later Singer W12.
1890's-1900, No13 & 14.
1890's-1905, No15, industrial flatbed tailoring machine.

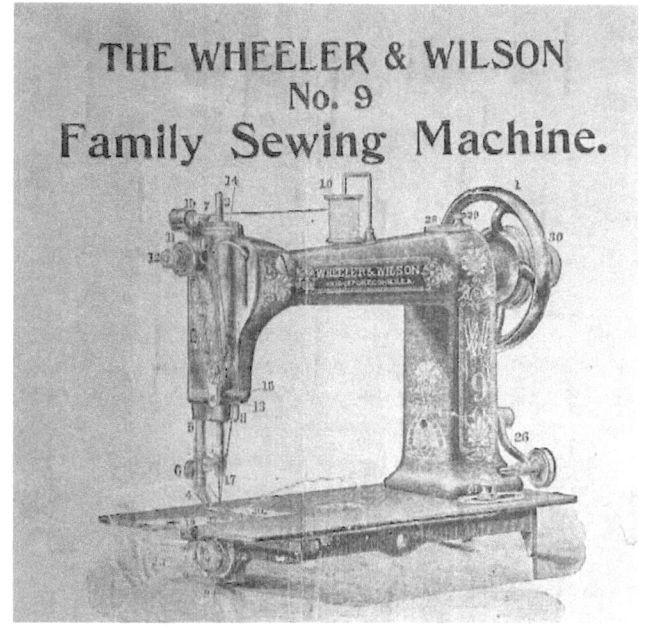

Singer continued making excellent sewing machines at the Bridgeport factory for many years. Initially they carried on with the same machines but slowly they used the best ideas from both companies to combine into a whole new range of machines. Before long the Singer 101, 201, 221 and more would be a combination of the best from both Singer and the Bridgeport engineers.

Chapter Sixteen

The end of an era

In the 1880's the large Wheeler & Wilson D10 was advertised as the fastest lockstitch in the world, capable of over 2,500 stitches per minute with a steam powered belt drive. It was offered with the normal teeth feed or for leather work with a roller or wheel feed. The machine also came with a drop feed and a large bobbin to hold more thread. It was for the industrial-factory market.

The Wheeler & Wilson D12

The Wheeler & Wilson D12 came out around 1892. It was a zigzag sewing machine. They are extremely rare today. A cam mechanism moved the top needlebar in time with the lower bobbin assembly. Two threads could be used on the top of the machine for extra strength. This machine was set flat into a table which became known in the trade as a 'flat-bed'. It was because of models like these that the Singer Company was desperate to gain control of W & W.

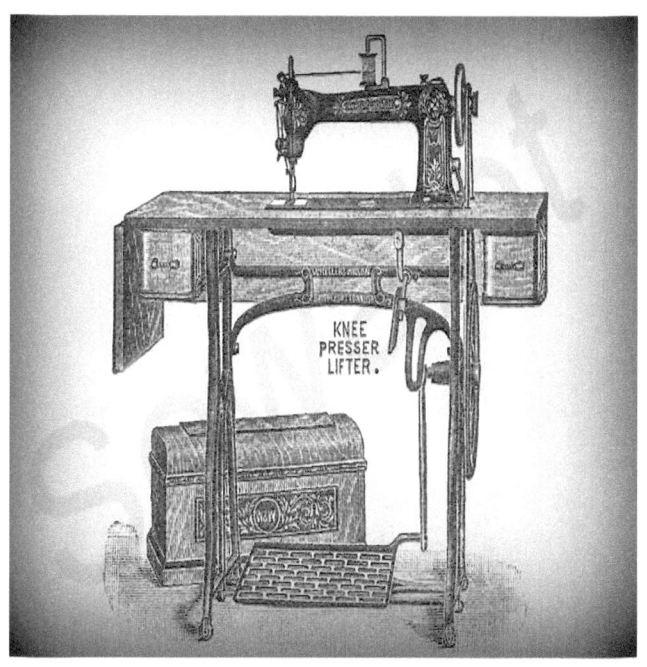

The W&W model N15 was one of their last models. It sold for $60 in 1904. A semi industrial available in electric as well, the machine was too late to save the old lumbering giant that was to fall prey to the jaws of the huge Singer Corporation in 1905. Image with kind permission of John Tierney.

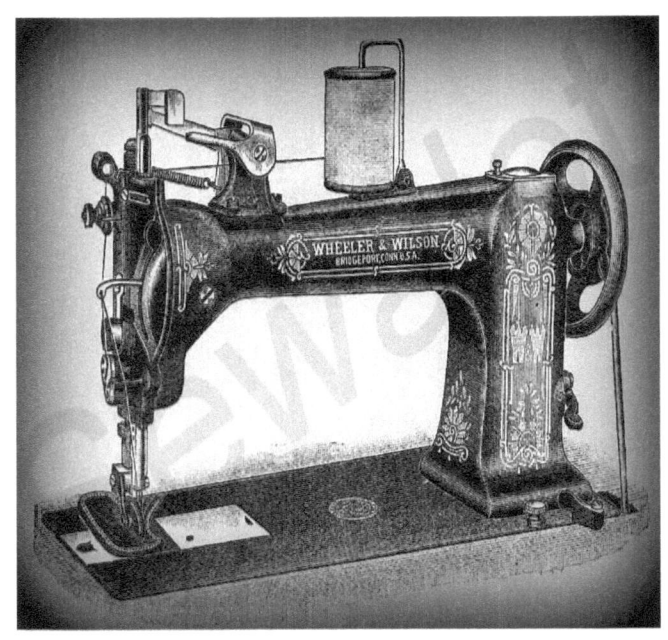

Wheeler & Wilson model N12
Wheeler & Wilson made this superb model N12 with a rotary feed wheel and cutter combined. It was ideal for glove makers and fine leather workers. Unfortunately by the time this was made, around 1900, W&W were in decline. This machine cost $60 new and was available on hire purchase at $4 per month. It was also offered in electric and barrel arm (see the 12W above). Image with kind permission from John Tierney.

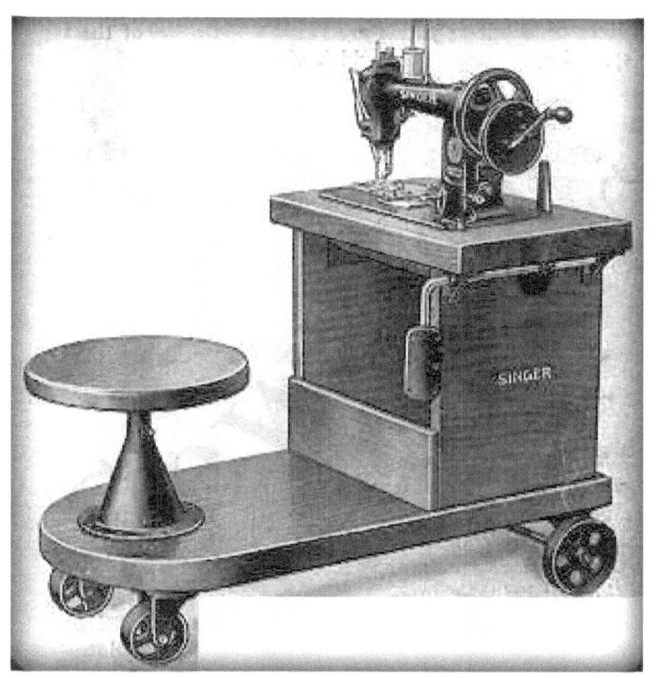

The Singer 11W was a Wheeler &Wilson heavy duty industrial available to the industry as electric, treadle or hand. Notice the knee lift and mobile work station on wheels for factory movement. Often used to sew the pockets onto denim where slow careful stitching was needed.

The Singer 12W came in many dresses including flatbed and, as we have here, the barrel arm. Again this is an industrial that was made at the Bridgeport Factory and sold for decades after W&W sold out. The Model 12 was advertised as the fastest lockstitch sewing machine in the world, offering over 3,000 stitches every 60 seconds. This could only be achieved because of the high standard of finishing to the machines. An average domestic sewing machine today still struggles to do more than a few hundred stitches per minute.

After purchasing Wheeler & Wilson, Singer restructured the sprawling factory.

They harmonised agents and sales to get the best of both worlds. They kept the takeover as quiet as possible, simply stating that they *'had assumed control and direction of the business.'*

This masterstroke helped to give Singer world dominance for decades.

Various models were manufactured at the Bridgeport factory and badged as Singer's but many were just earlier Wheeler & Wilson models with slight modifications.

Singer were also delighted to obtain Wheeler & Wilsons industrial machines, which they produced for decades. Many and are still in use today, especially in the jeans industry.

Although the Bridgeport factory shrunk in size continually, as demand dropped they made parts and industrial models right up until the late 1950s and beyond.

The Wheeler & Wilson in Singer dress circa 1909.

This cross over model is wearing Singer dress but still has the original Wheeler & Wilson badge in the bed. Were they using up the castings or was this beauty completely refurbished? A modern industrial MY1014 needle works well in the model 9.

The Singer 9W, a Wheeler & Wilson model D9 with Singer Sphinx decals circa 1912. Note the Singer still has the W&W brass badge in the bed! Using up old stock probably. These are the rarest of the cross-over Singer-Wheeler & Wilson machines, great for collectors if you can find one!

The Singer Company continued to manufacture and improve the model 9W for a number of years under its own brand name and even used several of its features on their models. It was a near perfect machine and so light to use it nearly sewed by itself. Even today the machine makes a perfect stitch, but needles are hard to come by.

A rare copy of W. A. Emery, foreman at the Lenoir Textile Mill in Loudon County. Probably taken around 1905 in Martel, Tennessee. After the Singer takeover he seems to be working as an agent or journeyman selling both Wheeler & Wilson and Singer machines. The treadle on his wagon is a Singer model.

It was only with the emergence of the Asian economies as a huge marketing and manufacturing force that the mighty Singer Company was eventually felled (in the 1980s). Singer survives today as a brand name.

Turn Drudgery into Pastime with Wheeler & Wilson. The Wheeler & Wilson No 9. All the advertising in the world and loads of awards could not save the company from Singer's takeover in 1905.

Bear rug! I love it. Not very PC today and what about fleas!

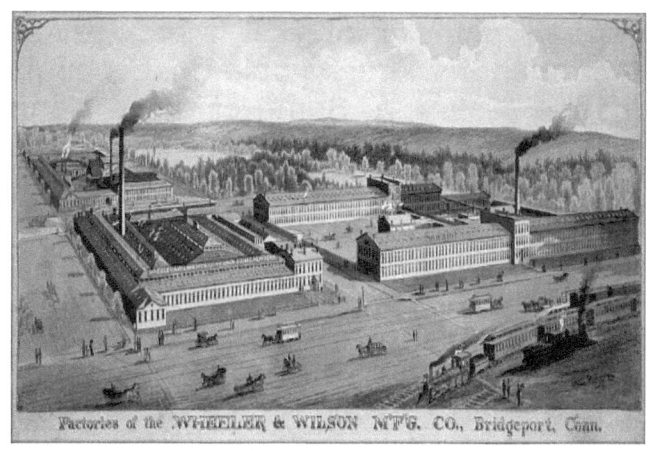

The huge Wheeler & Wilson factory which became a part of the Singer Company after the takeover in 1905.

Here you can see another of the Wheeler & Wilson machines that was rebadged and sold as the Singer 1W or 1W1. The machine had been so popular from the 1850s up until 1905 that Singer continued manufacturing the model for several years. Notice the small block on the lower bobbin-hook. That holds a brush that wipes away any thread. Only recently has that come back on several new sewing machines.

One final point of interest is that several other manufacturers copied the popular Wheeler & Wilson. Some of these clones were made under licence and some when the patents ran out. The biggest was probably the German giants Frister & Rossmann who made a near perfect replica of the early Wheeler & Wilson No1. Similar to the Singer model 1W.

A German Frister & Rossmann clone.

And so my friends we come to the end of our Wheeler & Wilson story. It has been great fun researching the famous sewing machine pioneers of the past.

Thirty years of compiling notes have been condensed into a few pages. I'm amazed no one has ever done it before.

More importantly, I do hope that you have found it interesting.

Allen B. Wilson and Nathaniel Wheeler were once involved with the largest sewing machine business in the world. Now they are little more than a small footnote in our history. However at one time Wheeler & Wilson were giants amongst the few. They were men who really did make a stitch in time!

The End

Wheeler & Wilson
'A Stitch in Time'
By
Alex Askaroff

To see Alex Askaroff's work
Visit Amazon

www.sewalot.com

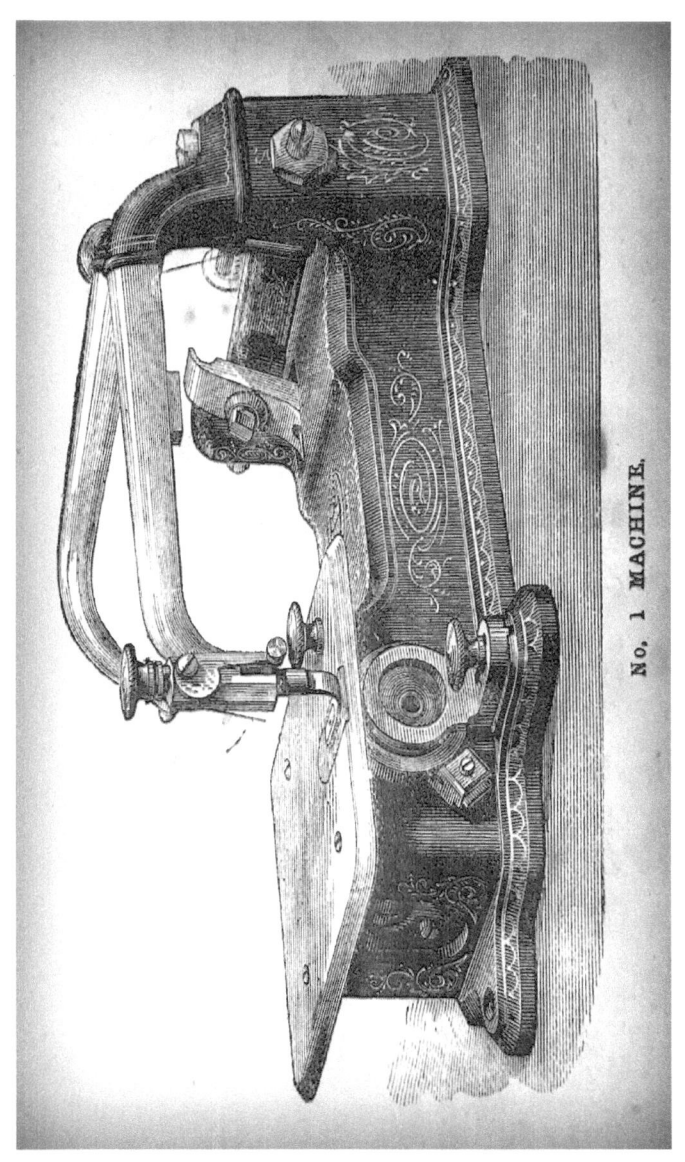

No. 1 MACHINE.

Isaac Singer
The First capitalist

Most of us know the name Singer but few are aware of his amazing life story, his rags to riches journey from a little runaway to one of the richest men of his age. The story of Isaac Merritt Singer will blow your mind, his wives and lovers his castles and palaces all built on the back of one of the greatest inventions of the 19th century. For the first time the most complete story of a forgotten giant is brought to you by Alex Askaroff.

On The Road Series

There are seven books in Alex Askaroff's **On The Road Series**. They cover his working life around Sussex encompassing a world of stories from the ages.

Book One: Patches of Heaven

Book Two: Skylark Country

Book Three: High Streets & Hedgerows

Book Four: Tales From The Coast

Book Five: Have I Got A Story For You

Book Six: Glory Days

Book Seven: Off The Beaten Track

"If you read any of Alex's 'On The Road Series' you will read them all. They are totally addictive, beautifully crafted and wonderfully inspiring."
Eliza Cooper

Alex Askaroff at Birling Gap

For collectors and enthusiasts
of antique sewing machines and great stories
why not visit

www.sewalot.com

For other publications
By
Alex Askaroff
Visit Amazon

No1 New Release. No1 Bestseller, Amazon certified.

If this isn't the perfect book it's close to it!
I'm on my third run through already.
Love it, love it, love it.
F. Watson USA

Patches of Heaven is the first book in Alex's popular 'On The Road Series'. We start Alex's working life and follow him as he earns his living. With Nine No1 New Releases on Amazon, Patches of Heaven with enthral readers of all ages.

Alex's second book in his hilarious 'On The Road Series' continues with his travels. We meet forgotten war heroes and crazy customers by the bucket load, from the 1930s debutant balls at Buckingham Palace to a sailor who had a lucky escape from HMS Hood, before its encounter with the formidable Bismarck Battleship.

Once again Alex has worked his magic. These true stories will have you in stitches, you may even shed a tear but you will be left with a happy heart.

Tales from the Coast, Book Four in Alex's On The Road Series, continues the true stories which he brings both England's history and people vividly to life. The stories are as pleasurable as a warm bath after a long day. From the disappearance of Lord Lucan in Uckfield to the Buxted Witch, from William Duke of Normandy to Queen Elizabeth's Eastbourne dressmaker, Tales from the Coast is crammed with a fascinating mix of true stories that will have you entranced from start to finish.

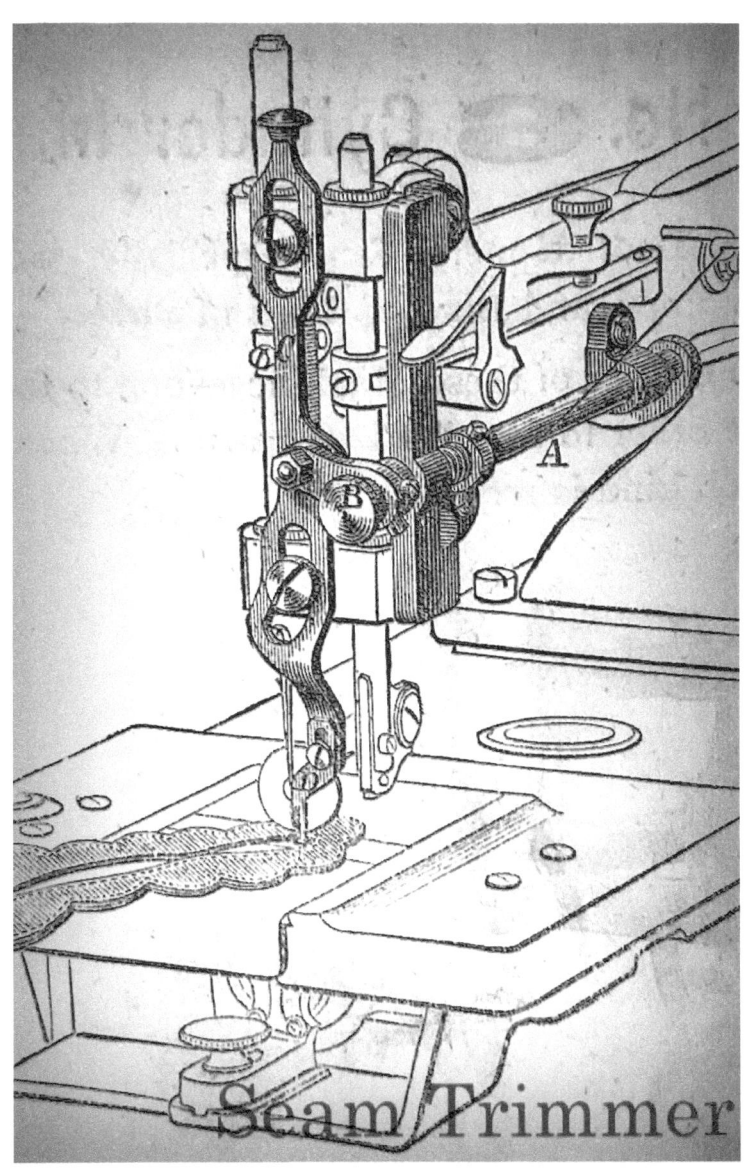

Yet again Alex has woven his magic. I kept saying I never knew that and I'm a local. This may just be one of the best books I've ever read!
J. Vincent

Alex, I've read every book James Herriot ever wrote, and my favorite topics in his books are about the animals and the meals, just like my favorite stories in your books are the ones that talk about your experiences working in people's homes. I love them.
Thank you so much.
Joe Edmiston
Louisville, KY

Alex Askaroff has had Nine No1 New Releases on Amazon. For decades Alex has been enthralling readers around the world with his writing. Off The Beaten Track is the seventh and final book in his 'On The Road Series' and completes his working life before retirement.

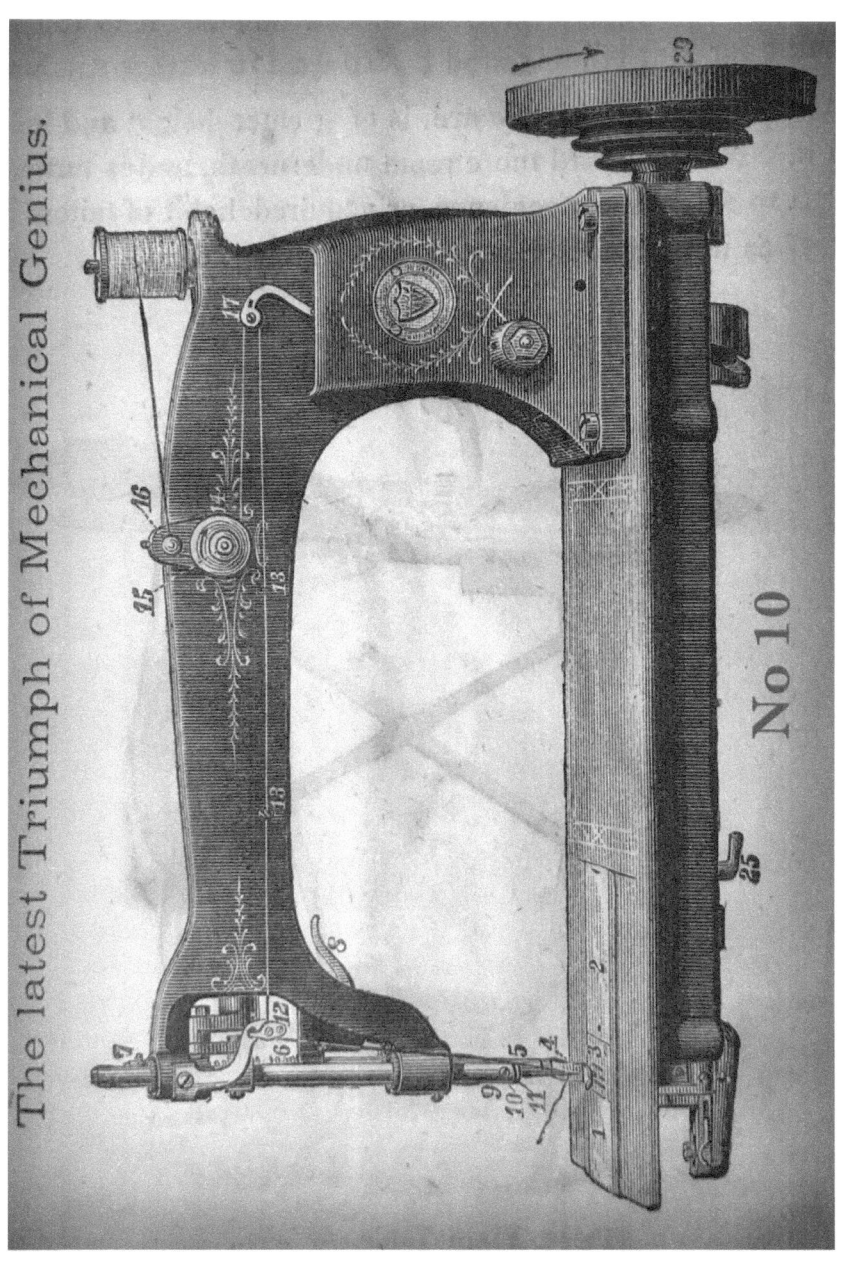

The 7th Duke of Devonshire's dream started in the middle of the Victorian Era and continues to flourish to this day. World renowned author Alex Askaroff tells the story of Eastbourne in his own unique style, reviving long forgotten characters from the town. We meet local families, fishermen, smugglers, kings and queens, ghosts and even an old witch. It is a tale not to miss.

www.sewalot.com
Alex's No1 antique sewing machine site.

Sir Sewalot, protector of Sewalot.com
Sewalot is where I publish histories of the pioneers that never made it into my books. From A F Johnson who fought so bitterly with Edward Allen Gibbs for the rights to the revolving hook, to George Washington Gates, the only sewing machine maker in Toronto, Canada.

Elias Howe
The Man Who Changed The World
No1 New Release Amazon Oct 2019.

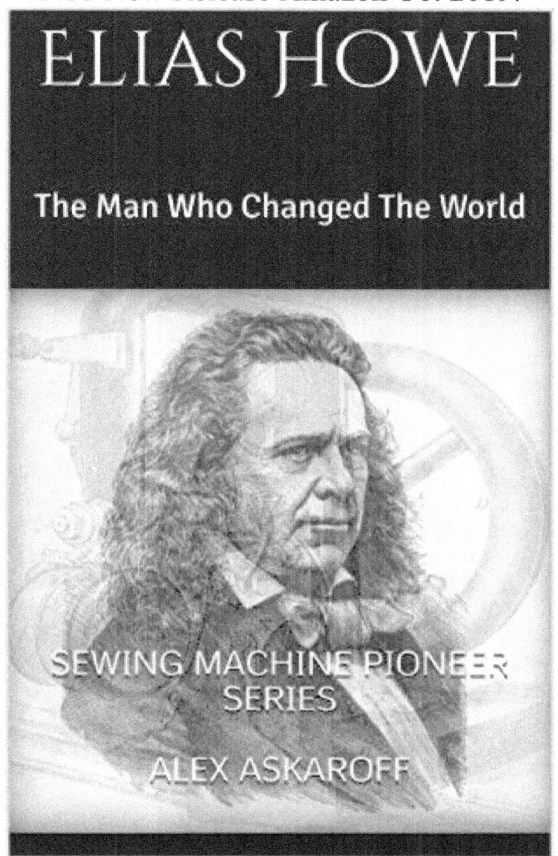

Anyone who uses a sewing machine today has one person to thank, Elias Howe. He was the young farmer with a weak body who figured it out. Elias's life was short and hard, from the largest court cases in legal history to his adventures in the American Civil War. He carved out a name that will live forever. Elias was 48 when he died. In that short time he really was the man who changed the world.

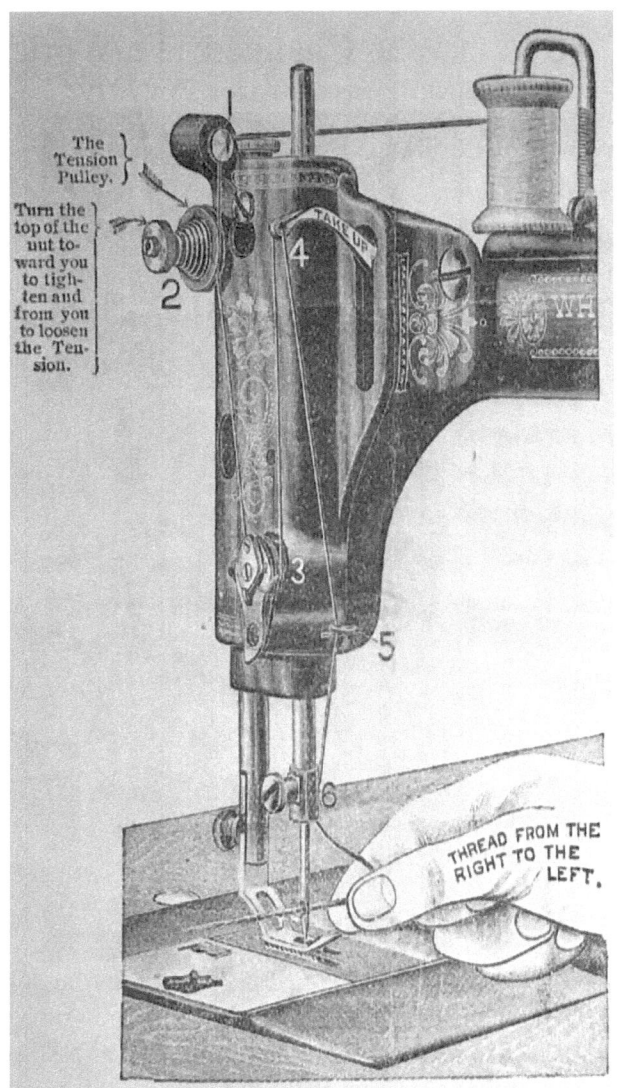

Notice how on the W&W 9 the needle is threaded. The Singer 201 which came out a few years later had the same needle threading. The machine used a 12x1 needle, a modern equivalent is the MY 1014.

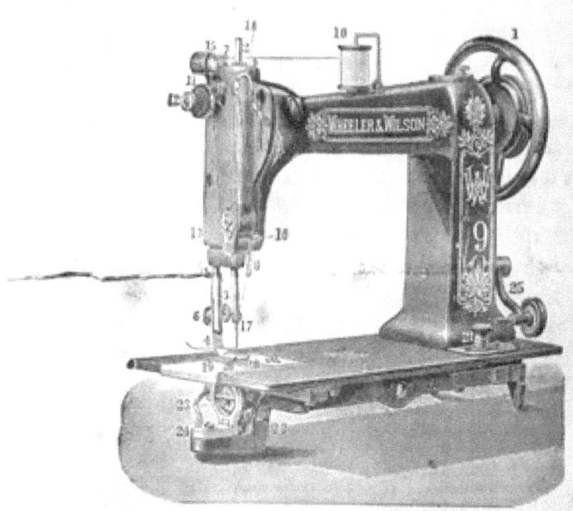

To Thread the Machine. Place the spool on the spool wire, pass the thread through the thread check (15), down in front, in the groove, and once around the tension pulley (11), thence under the thread guide (13), through the take-up (14), the thread leader (16), and the needle bar thread guide (17) to the needle. (See that the thread is pulled into the thread guide until it passes both little hooks, and within the bend of the controller spring.) Then thread the needle from right to left, all as seen in the cuts.

Let the thread extend through the needle two or three inches when the take-up is fully up. Now hold the end of the upper thread loosely between the thumb and finger, and turn the hand-wheel carefully from you while the needle goes down and up once, then draw the upper thread and with it the under thread up through the hole in the throat-plate (18), and pass them both back under the presser (4).

To Commence Sewing. Place the cloth beneath the needle, and let down the presser-foot by dropping the presser-lifter. Place your finger upon the ends of the threads, till one or two stitches are taken. Start the machine by turning the hand-wheel from you. Do not pull or push the work; the machine feeds it. If it does not, the feed is shut off. If the machine is accidentally turned backward no injury will be done, and it is only necessary to start again in the right direction and continue the sewing.

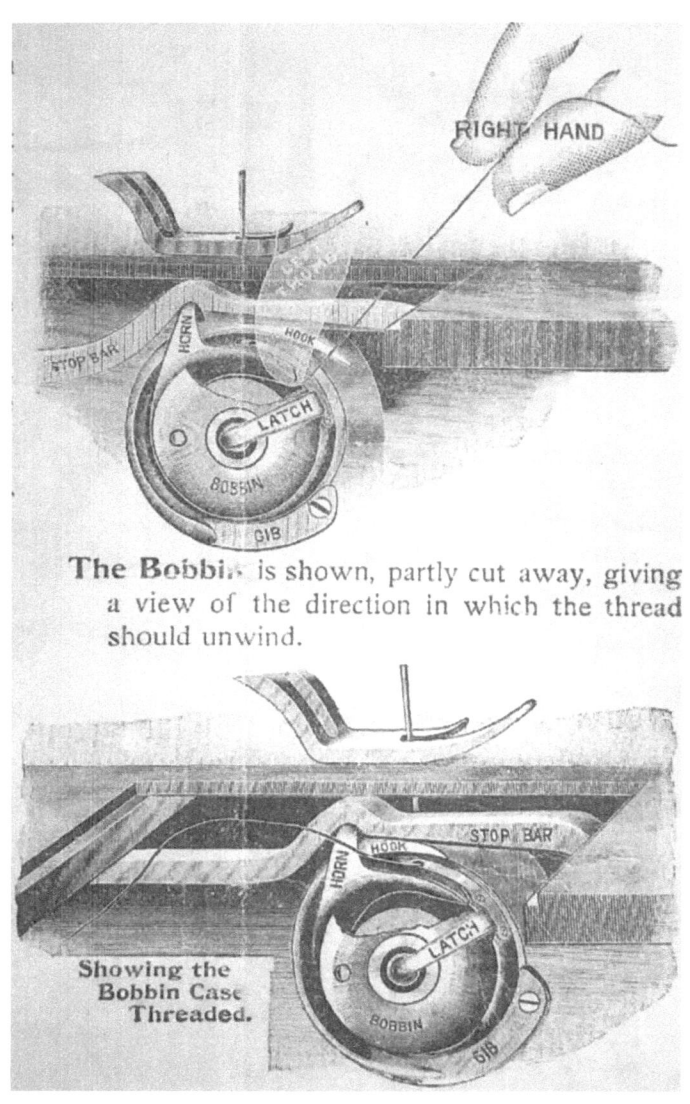

The Bobbin is shown, partly cut away, giving a view of the direction in which the thread should unwind.

Showing the Bobbin Case Threaded.

Wheeler & Wilson

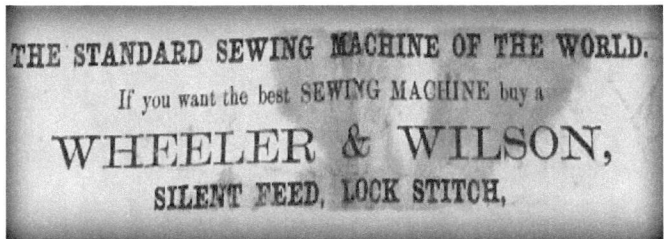

From their start in the 1850s Wheeler & Wilson sewing machines grew at a phenomenal rate. During the years of the American Civil War they produced more sewing machines than all the other makers on the planet combined.

Along with Singers, Wheeler & Wilson dominated the 19th Century, producing some of the finest sewing machines that money could buy. For 50 years they became the 'sewing machine of the world'.

www.ingramcontent.com/pod-product-compliance
Lightning Source LLC
Chambersburg PA
CBHW021417210526
45463CB00001B/407